THE
SCIENCE OF GOLF

THE
SCIENCE OF GOLF

JOHN WESSON

OXFORD
UNIVERSITY PRESS

OXFORD

UNIVERSITY PRESS

Great Clarendon Street, Oxford OX2 6DP

Oxford University Press is a department of the University of Oxford.
It furthers the University's objective of excellence in research, scholarship,
and education by publishing worldwide in

Oxford New York

Auckland Cape Town Dar es Salaam Hong Kong Karachi
Kuala Lumpur Madrid Melbourne Mexico City Nairobi
New Delhi Shanghai Taipei Toronto

With offices in

Argentina Austria Brazil Chile Czech Republic France Greece
Guatemala Hungary Italy Japan Poland Portugal Singapore
South Korea Switzerland Thailand Turkey Ukraine Vietnam

Oxford is a registered trade mark of Oxford University Press
in the UK and in certain other countries

Published in the United States
by Oxford University Press Inc., New York

British Library Cataloguing in Publication Data
Data available

Library of Congress Cataloging in Publication Data
Data available

Printed in Great Britain
on acid-free paper by
CPI Antony Rowe, Chippenham, Wilts

ISBN 978–0–19–922620–7 (Hbk)

10 9 8 7 6 5 4 3 2 1

CONTENTS

ACKNOWLEDGEMENTS

First I must thank my wife, Olive, not only for her patience during the writing of this book but also for her help with numerous experiments and the compilation of statistics.

I am particularly grateful to Lynda Lee and Stuart Morris. Lynda typed the manuscript and cheerfully dealt with the large number of modifications that arose as the text developed. Stuart drew all of the figures, about 200, with his usual skill and attention to detail.

I needed the help of a professional golfer to carry out basic experiments on the swing and several other subjects dealt with in the book. I was therefore fortunate to have the enthusiastic collaboration of Ian Mitchell in these experiments, together with his advice on their interpretation. I am also grateful to David Goodall who carried out the associated filming.

The comprehensive series of experiments described in Chapter 13 was made possible by the collaboration of some of my golfing friends, and I would like to thank the participants: Jack Atkinson, Tony Davey, Peter Frearson, Eddie Lennon, Steve Lowman, Ian Mitchell, Peter Mitchell, Gwyn Morgan, Chris Parslow, Peter Sanderson, Adrian Smith, Mike Sumner, and Phil White.

Since the book is based almost entirely on new material I was very lucky that several of my golfing and scientific colleagues were willing to read all, or parts, of the manuscript to identify errors and make suggestions for improvements. In particular I would like to thank Barry Alper, Jim Hastie, Ron Howarth, John Maple, Bob McLaughlin, Brian Payne, Robin Prentice,

Francis Sabathier, and Bert Shergold. I am especially grateful to Tim Luce who read and commented on the sections dealing with American golf, with which I am less familiar. I would also like to thank Trevor Jenkins who, although not a golfer, was willing to read the whole of the manuscript. His eagle-eye picked out typographical errors that I had missed, together with some careless punctuation.

I have benefited from discussions with experts around the world and from information provided by many individuals and organizations. I would particularly like to thank the following: Steve Aoyama of Acushnet for information on golf balls, John Barton of *Golf Digest* for advice on the number of golfers around the world, Alan Clayton for discussions on the mechanics of clubs and balls, Caroline Capocci of the General Register Office for Scotland for data on the growth of Scotland's population, Lawrence Donegan of the *Guardian* for help in finding golf statistics, Raymond Penner for helpful discussions on the physics, Alan Sykes for help with the high-speed photographs, Karen Wesson for advice on physiology, David Wesson for guidance on computing, the PGA Tour who supplied me with a great deal of information, the Ladies PGA for data on the PGA prize purse, and the National Golf Foundation for statistics on the growth of the number of US golf facilities.

I am also grateful to Eddie Lennon and Gerald Mace for providing statistics on players' scores and to the Drayton Park Golf Club, Stephen Styles and the Frilford Heath Golf Club, and Adrian Smith and the Hadden Hill Golf Club for providing further statistics.

My special thanks go to Mike Morley, who kindly allowed me to use the facilities of the Hadden Hill golf course for experiments and has generously given of his time for discussions on a variety of issues concerning the economics of golf courses.

PREFACE

There are hundreds of books explaining how to play golf, and this book is not one of them. We are concerned here with the science, rather than the art, of playing golf.

It is quite understandable when people ask—what *is* the science of golf?—because it is not immediately obvious. The reason is that much of what happens in golf is not seen directly by the players. For example, the impact of the club on the ball occurs in less than a thousandth of a second and this is so brief that a proper understanding of hits and mis-hits has to come from physics. Again, we cannot see the airflow over a ball in flight and to understand the flow and how drag, spin, and wind affect the range, we have to turn to aerodynamics.

However, the mechanics of the game is only part of our subject. Two chapters of the book examine the main handicap systems and their implications for the players in both matches and competitions. Three further chapters then discuss the performance of players, the equipment of golf, and the economics of the game.

It is the nature of science that one question leads to another, and so no account is ever complete. The same is true here and I hope that readers will find some pleasure in discovering and thinking through such further questions.

John Wesson
January 2008

THE
SCIENCE
OF GOLF

There are two basic reasons why golf is such a splendid game. The first is the variety of situations players face as they make their way from tee to hole. The second is the handicap system, which allows players with widely different abilities to compete with each other. When we attempt to understand these matters we soon find that we are involved in scientific issues.

The first thing a beginner learns is that it is not so easy to hit a straight drive, and that ill-directed shots can lead to the challenges presented by rough grass, trees, ditches, bunkers, lakes, and 'out of bounds'.

But why didn't the ball go straight in the first place? It could be that the swing of the club was out of line. However, when we analyse the mechanics of this type of mis-hit we find it is very forgiving, in that the angle of the ball's departure is much less than the out-of-line angle of the swing. So that is encouraging. Unfortunately, the same analysis tells us that the out-of-line swing also imparts a side-spin to the ball, and it has been known since the seventeenth century that spin produce a sideways force on the ball. This is the source of the infamous slice. If we want to understand why the spin leads to a force we have to look at the fascinating subject of the ball's aerodynamics and understand the complex way in which the air flows round the ball.

To find out how much deflection the spin-force produces we need to calculate the ball's trajectory—and this introduces further issues, such as the air-drag on the ball. In the following chapters we shall look at all of these matters.

However, this discussion has taken us too far ahead— let us go back to the beginning. The first subject we shall deal with is the swing. This is usually examined in the context of improving the player's technique, but the purpose here is not to advise how to make the swing, it is to discover what happens when we do. So, in the next chapter, we shall analyse the club's motion during the swing and determine the force and power required in accelerating the club through to impact with the ball. To further understand the swing we shall look at the bending of the shaft and its influence on the effective loft of the club.

The analysis of the swing rests on the use of Newton's second law of motion and we shall find that, throughout the physics chapters of the book, Newton's laws play a central role. However, there are parts of the subject that have to be treated empirically and this is done by introducing a variety of coefficients. When the ball bounces, off the ground or off the clubface, its 'bounciness' is measured by the coefficient of restitution. When the ball slides or rolls, the slowing is measured by the coefficients of sliding and rolling friction. And in the ball's flight through the air its motion is partly determined by the drag and lift coefficients.

The third chapter is concerned with the impact of the club on the ball and finds that there is a lot of physics involved in the brief half-thousandth of a second of contact. It is here that we shall examine how the theory of the impact explains the consequences of the various types of mis-hit.

The fourth and fifth chapters deal with the aerodynamics of the ball. The understanding of the aerodynamics follows from the scientific contributions of many scientists, including Newton, Robins, Bernoulli, D'Alembert, Magnus, Stokes, and Prandtl. It is interesting, however, that the importance of dimples was discovered by the golfers themselves.

The science described in these chapters puts us in a position to study the trajectory of the ball from a specified hit and to calculate its range. Given the club-head speed and loft of a driver, we can calculate the speed, spin, and launch angle of the ball, and these in turn determine its trajectory and range. The theoretical procedure is outlined in Chapter 6 and the range is calculated for a variety of cases in Chapter 7. From this we come to learn the optimum loft for a given player and ask how sensitive the achievable range is to the choice of loft.

One of the uncertainties in calculating the range is the run of the ball. The trajectory calculations give the carry, which is the horizontal distance the ball travels before reaching the ground. The subsequent run of the ball presents a problem because the bounce and roll of the ball depend so much on the ground conditions. In the range calculations these complications are avoided by considering 'typical' cases, representing average conditions. However, because bouncing and rolling are interesting in their own right and because of their importance on the green, Chapter 8 is devoted to an analysis of these subjects.

We then come to putting, the subject of Chapter 9. As in a drive, the striking of the ball is important, but now there is the further complication of slopes, to say nothing of the effect of winds and the vagaries of hitting the hole. Add to this the imperfections of the ball and there is a lot to understand. And we shall see why it is useful to put golf balls in salt water.

A question which often arises is—what is the probability of a 'hole-in-one'? There is, of course, no simple answer. The probability depends on both the player and the hole, and so there are many answers. However, it is possible to make a reasonable assessment of how the probability depends on the player's handicap and the length of the hole, and the results are given in a short Chapter 10.

At the beginning of this introduction the importance of handicaps in the success of the sport of golf was recognized. The author's experience is that the role of handicaps is widely misunderstood. Most players appear to believe that the purpose and effect of handicaps is to give all players an equal chance of winning. Whatever the intentions of the designers of the British handicap system were, equality is not what it provides. In the American handicap system a small inequality is explicitly designed into the procedure for calculating handicaps. Chapter 11 describes the handicap systems and examines their implications.

Chapter 12 is devoted to matches and competitions. If the handicap systems gave players truly equal chances then the theory of competitions would be of no interest. In matches each player or pair would be equally likely to win and in competitions all entrants would have an equal chance. The actual situation is different and very interesting. We have to take account not only of the advantage or disadvantage that the handicap confers directly on players with different handicaps but also of the variability of their scores, which also depends on their handicaps. When these consequences of the handicap systems have been calculated, readers can judge for themselves whether the inherent inequalities are justified.

There are estimated to be about 60 million people who play golf. Most of these are casual golfers who play a few games a year. The number of regular golfers who belong to a club and have a handicap is probably more like 20 million. In Chapter 13 we look at these golfers and examine the distribution of their handicaps. Combining this with the results from professional tournaments, an attempt is made to produce a distribution of abilities across the whole range from the high handicappers through to Tiger Woods.

Also in Chapter 13 we look at the games of players and ask what aspects of their play are important in determining their ability, as measured, say, by their handicaps. To investigate this, a series of experiments was carried out on the course with a group of volunteer players. The results enable us to assess, for example, the relative importance of long driving as compared with putting ability.

The development of golf equipment, essentially clubs and balls, has been driven by two factors—science and technology. The scientific contribution is illustrated by the improved understanding of the flight of the balls and by the analysis of the effect of clubhead geometry during impact with the ball. Technologically, balls have been improved by the use of new materials, through leather-encased feathers to gutta-percha and on to the two-piece ball with a rubber core and resilient synthetic cover. Alongside these improvements has been the development of techniques for the production of quality balls on a massive scale.

In the case of the clubs there was a gradual improvement in the development of wooden clubs, finally leading to the production of a sophisticated design and a product that was a work of art. The introduction of steel clubs with their hollow construction allowed a greater flexibility in the distribution of the weight of the club and the design of the modern large-headed drivers. A further factor has been the introduction of carbon shafts, which gives players a choice between carbon and steel. These developments are described in Chapter 14, which also addresses the question of how much the improved performance of players is due to the improvements in the equipment.

The final chapter deals with the economics of the game. The aspect which, through television, receives most attention is the professional game. There are now many 'Tours' which allow the top players to derive a considerable income, and we shall see how the tours have grown from comparatively humble beginnings to their present affluent state.

However, financially the professional game is dwarfed by the enormous expenditure of amateur players, partly on equipment but predominately in the payment of club membership fees and course fees. We shall look at the economics of golf clubs, both in terms of the historic growth in the number of clubs and through their need to attract sufficient players, in competition with their neighbouring clubs, in order to cover their costs.

2 THE SWING

Ben Hogan said that a good golf swing requires twelve unnatural movements. Given the dozens of pieces of advice offered in 'How to play golf' books, he might have oversimplified the problem. At first sight this does not seem fertile ground for science, which is at its best when it is possible to identify the key issue and neglect the inessential features.

There is no doubt that advice from a good coach can transform a player's swing, but it is based mainly on the insights acquired by good golfers over more than a hundred years. It is easy to give plausible explanations as to how the suggested changes work through to a successful result, but rather difficult to demonstrate the relation scientifically. The consequences of changing the angle of the thumb on the club or the movement of the shoulders, for example, do not present a straightforward problem for physics. So, while science can help with insights, the player-centred approach is more of an art than a science.

However, it is possible to look at the matter another way. We can start from the observed motion of the club, which is known quite accurately from photographic evidence, and analyse this motion to determine the forces which bring it about. This way we come to see clearly the sequence of events during the swing. In turn, the forces on the club imply forces on the player, through to his hands and arms to his body, and so the analysis of the behaviour of the club can be connected to the experience of the player as he makes the swing.

The speed of the clubhead in a long drive typically reaches around 100 miles per hour at the time of impact with the ball, and the distance the ball is hit increases by about 3 yards for each 1 mile per hour increase in clubhead speed. So clearly the purpose of the drive is to produce maximum possible speed of the clubhead without, of course, any significant loss of control.

The control of the clubhead achieved during the swing is remarkable. The starting point is the address of the ball, placing the clubhead at just the position to which the golfer hopes it will return at high speed about a second later. The mind and body register this starting position and the clubhead is then taken on the backswing. The length of this journey depends on the confidence of the golfer, less-experienced golfers often preferring a shorter swing, but a typical length is 14 feet. The clubhead is then brought forward on its return path. That is another 14 feet, making 28 feet in all. Amazingly, the clubhead now hits the ball within, say, half an inch of the centre of the clubface. A required half-inch accuracy after a journey of 28 feet! It is probably best to dismiss such thoughts when you approach the swing.

The double pendulum model

The movements involved in the swing are extremely complicated, but fortunately the movement of the club and the arms can be represented by a simple model. Imagine a pendulum which turns about its hinge, and then add a second pendulum, attached and hinged at the end of the first. In this double pendulum model, the first pendulum represents the arms and the second represents the club.

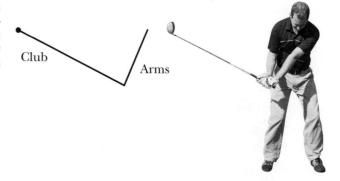

Fig. 2.1. The double pendulum model. The upper limb of the pendulum represents the arms, hinged between the shoulders, and the lower limb represents the club, hinged at the wrists.

Club

Arms

This is illustrated in Figure 2.1, which shows how the double pendulum relates to the actual golfer. The first pendulum, which is the arms, pivots about a point between the shoulders. The second, club pendulum, pivots about the wrists.

The motions of this model can be described by mathematical equations, which are based on Newton's famous second law of motion.

$$\text{Force} = \text{mass} \times \text{acceleration}.$$

However, these equations are very complicated and in themselves add little to our understanding. Nevertheless, it is encouraging that, when they are solved numerically, they give a good representation of the real motion of the club and arms. Here, however, we shall rely on direct observation of the swing.

Our basic swing

The line of the swing lies almost in a plane, typically at an angle around 45° to the vertical, as illustrated in Figure 2.2. The swing is best recorded photographically by viewing this plane at a right angle, and when we use the double pendulum model it will be displayed as viewed at this angle.

Fig. 2.2. The path of the club lies approximately in a plane.

Each player, of course, has his personal swing, and even this varies from shot to shot. However, it will simplify our approach if we focus on one typical swing. It will have all the elements for understanding the behaviour, and other swings, shorter or longer, faster or slower, will have the same essential features.

The details of this basic swing were recorded using a cine camera to film the swing of a professional golfer. The result is shown in Figure 2.3 where the swing is represented using the double pendulum model, the position of the club and arms being shown at intervals of 0.02 seconds. The forward swing takes just over a quarter of a second and produces a clubhead speed of just over 100 miles per hour. The resulting drive reached 270 yards.

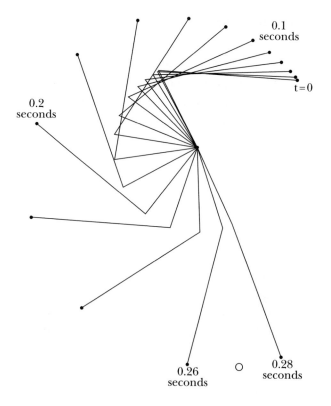

Fig. 2.3. The motion of the club and arms derived from cinephotography of a swing, and represented by the double pendulum model.

Two features are obvious from the diagram. First, unsurprisingly, the speed of the clubhead increases throughout the swing. Second, the clubshaft initially trails behind the arms and the large angle between them persists well into the swing, after which the clubhead moves quickly to 'catch up' with the arms.

The clubhead speed

The first information we can obtain from Figure 2.3 is the speed of the clubhead during the swing. By measuring the distance moved by the clubhead in each interval of time we can calculate its average speed during that interval, as shown in Figure 2.4.

When we have done this for all of the time intervals, we can draw a graph of speed against time and this is given in Figure 2.5. We see that the speed increases at a comparatively slow rate in the first half of the swing and then accelerates at a higher rate to reach 107 miles per hour. The clubhead then loses momentum on impact with the ball and slows down thereafter.

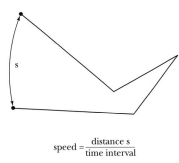

$$\text{speed} = \frac{\text{distance } s}{\text{time interval}}$$

Fig. 2.4. The speed of the clubhead is determined by measuring the distance, s, it moves in a given time interval.

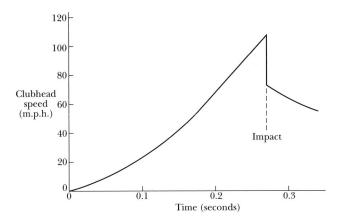

Fig. 2.5. Clubhead speed during the swing.

Accelerations and forces

Now that we have the time dependence of the speed, it is straightforward to calculate the acceleration of the clubhead along its path. This acceleration is just the rate of change of the speed, and the acceleration at any given time is simply given by the slope of the graph in Figure 2.5 at that time. Knowing the acceleration we can then calculate the associated force on the clubhead using Newton's second law. To obtain the force we just have to multiply the acceleration by the mass of the clubhead.

Before proceeding to these calculations it is important to clarify two matters. First, when the calculations are carried out it is necessary to use a consistent set of units. Scientists use the Système International, SI, units and in the case we are studying, the acceleration is then in metres per second per second, the mass is in kilograms, and the resulting force is given in newtons. However, for most golfers these units are unfamiliar. Here, and throughout the book, the reader is asked to take such calculations as done and accept the answers in more familiar units.

The second point is that forces will be expressed in pounds. Now the pound is actually a unit of mass and when a force is given in pounds it is a short way of saying that it is equal to the weight of a mass of that number of pounds, where the weight, of course, is simply the force of gravity on that mass. It is confusing that the same name is used for both mass and weight. However, most golfers are familiar with a pound weight and have an intuitive understanding of its magnitude.

We can now proceed to the calculation of the acceleration from the slope of graph of the speed against time in Figure 2.5, and obtain the associated force

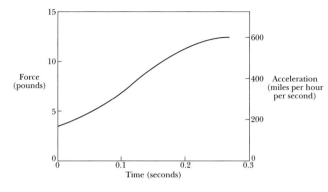

Fig. 2.6. The acceleration of the clubhead is calculated from the change of its speed during the swing. Multiplication by the clubhead mass then gives the driving force on the clubhead.

on the clubhead. Assuming a clubhead weight of 0.45 pound we obtain the graph shown in Figure 2.6.

We see that the acceleration rises to the quite remarkable value of 600 miles per hour per second. Our cars typically have an acceleration of a few miles per hour per second. The force on the clubhead that produces the large acceleration reaches about 12 pounds, which is 27 times the weight of the clubhead. It follows that the effect of gravity on the clubhead is comparatively small and can be conveniently neglected.

Figure 2.6, which gives the driving force accelerating the clubhead along its path, is just the beginning of the story. We shall discover that there is a much larger force on the clubhead, the centrifugal force. This force often causes some confusion and we shall digress briefly to clarify the subject.

Centrifugal forces

Let us take the simple example of a stone swung in a circle on the end of a string. Newton's laws tell us that the stone would move in a straight line if it was not acted on by any force. In our example, the cause of the change from motion in a straight line to motion in a circle is the force applied to the stone from the tension in the string as illustrated in Figure 2.7(a).

The acceleration towards the centre of the rotation is called the centripetal acceleration and inward force is called the centripetal force.

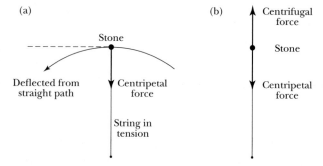

Fig. 2.7. Two ways of viewing the forces involved in curved motion. When a stone is swung in a circle at the end of a string the stone is pulled from a straight line path by the centripetal force supplied by the tension in the string. But as 'seen by the stone' the centripetal force balances the outward centrifugal force.

However, an equally correct description is obtained if we consider the behaviour from 'the stone's point of view' or, more formally, in a frame rotating with the stone. Seen in this frame, the stone is subject to two forces, the inward centripetal force and outward centrifugal force as shown in Figure 2.7(b). These forces are exactly equal and opposite, cancelling each other out and leaving no resultant force on the stone. As a result the stone maintains a constant distance from the centre of the circle.

The description in terms of a centrifugal force is more in keeping with our intuition. We are familiar, for example, with the centrifugal force we feel in a car which takes a bend sharply, or experienced more dramatically in some fairground rides.

Centrifugal force on the clubhead

The centrifugal force on the clubhead is given by the equation,

$$\text{Force} = \frac{\text{mass of clubhead} \times (\text{clubhead speed})^2}{\text{radius of curvature of path}}.$$

It is seen that the force varies as the square of the club-head speed so that, for example, doubling the speed produces four times the force.

We know the clubhead mass and have already cal-culated the clubhead speed for our chosen swing, so to calculate the centrifugal force all we now need to do is to measure the radius of curvature from the basic diagram in Figure 2.3. The procedure for calculating the radius of curvature is illustrated in Figure 2.8.

Two adjacent lines are drawn perpendicular to the clubhead's path and these lines meet at the centre of curvature, allowing a straightforward measurement of the radius of curvature. Of course, both the centre of curvature and the radius of curvature change as the clubhead moves along its path. These changes intro-duce some further effects but they are small and can be safely ignored. We now have a procedure that allows us to calculate the centrifugal force on the clubhead and its variation throughout the swing, and the result is shown in Figure 2.9. It is seen that at impact with the ball the centrifugal force has reached 60 pounds.

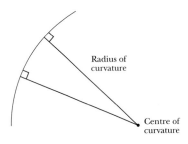

Fig. 2.8. Illustrating the procedure for determining the radius of the path of the clubhead at times during the swing.

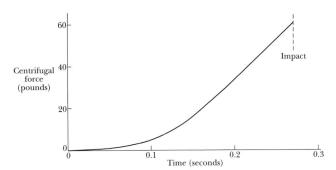

Fig. 2.9. Graph of the centrifugal force showing how it increases during the swing.

It is interesting to compare the centrifugal force with the driving force accelerating the clubhead along its path, which was given in Figure 2.6, and they are put together in Figure 2.10. We see that for almost half of the swing the driving force is larger than the centrifugal force. After that the centrifugal force comes to dominate, rising to a value which is 5 times as large as the driving force.

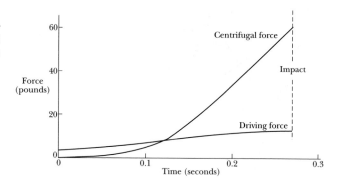

Fig. 2.10. Comparison of the centrifugal force on the clubhead and the driving force accelerating the clubhead.

The addition of forces

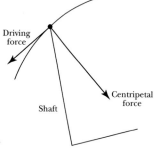

Fig. 2.11. The forces applied to the clubhead by the shaft.

We have now calculated the forces on the clubhead but we would like to know what this means for the player. The first step is to understand the force applied to the clubhead by the shaft. To do this we must be able to add the component forces on the clubhead to obtain the total force. In our case there are two forces from the shaft, the driving force along the path of the clubhead and the centripetal force, which is equal and opposite to the centrifugal force we have already calculated. These two forces are perpendicular to each other, as shown in Figure 2.11.

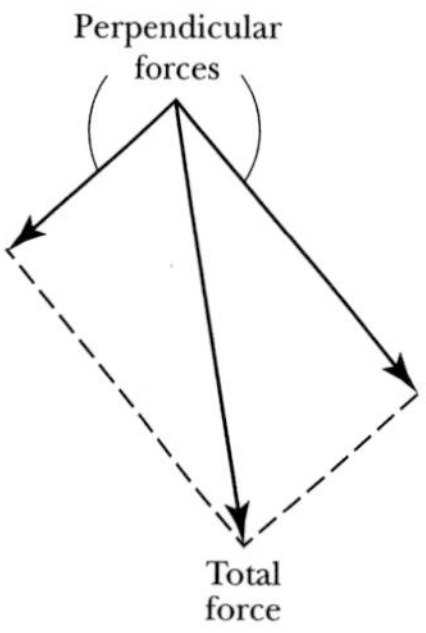

The rule for adding the forces is a simple one. We represent the individual forces by arrows whose length is proportional to their strength and whose direction gives that of the force. Since our two forces are perpendicular to each other they add as shown in Figure 2.12. The two forces give the sides of a rectangle, and the magnitude and direction of the total force is then given by the arrow on the diagonal.

Fig. 2.12. The addition of two perpendicular forces to give the total force.

Forces on the clubhead from the shaft

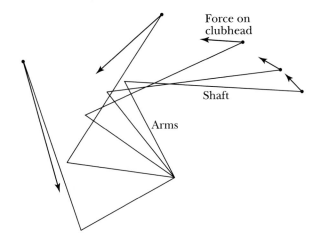

Force on
clubhead

Shaft

Arms

Fig. 2.13. Showing the magnitude and direction of the force on the clubhead during the early part of the swing, the force becoming increasingly aligned with the shaft.

To obtain the total force acting on the clubhead from the shaft we need to add the centripetal force to the driving force using the rule described above. Figure 2.13 shows the result at a sequence of times during the first part of the swing, the arrows giving the magnitude and direction of the force. At early times the force is at a substantial angle to the line of the shaft, initially about 50°. At this stage the force on the clubhead can be thought of as part pushing and part pulling the clubhead along its path. As the centrifugal force grows, the balancing centripetal force comes to dominate and the angle between the force on the clubhead and the line of the shaft decreases. Towards the end of the swing the force on the clubhead is along the shaft, so at these later times the shaft is essentially being pulled along its length. The angle between the shaft and the path of the clubhead is such that the force along the shaft balances the centrifugal force and also allows the clubhead to be pulled along its path, as shown in Figure 2.14. The graph in Figure 2.15 shows the variation of the angle between the force on the clubhead and the line of the shaft throughout the swing.

Fig. 2.14. At later times the force on the clubhead is essentially along the shaft. This force principally balances the centrifugal force but also accelerates the clubhead along its path.

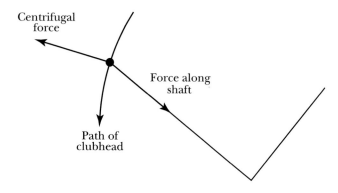

Fig. 2.15. Graph showing how the angle between the force on the clubhead and the line of the shaft changes during the swing.

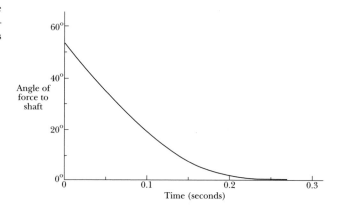

The role of the wrists

Throughout most of the swing, the pull along the shaft provides the dominant force on the clubhead. This shows that the wrists are mainly acting as a hinge and are not providing a substantial torque, which would be transmitted to the clubhead as a force perpendicular to the shaft. However, in the initial stage of the swing we find that more is required of the wrists. Figure 2.16(a) shows the situation at an early time when the angle between the arms and the shaft is 60°.

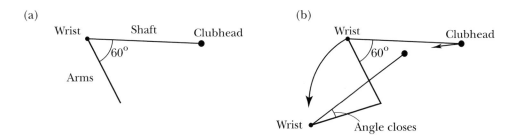

If the wrists were completely relaxed and acted simply as a pivot, the clubhead would 'jackknife', closing the angle between the arm and the shaft as illustrated in Figure 2.16(b). But there is a physiological limit on how much the wrists can be cocked, and during the early stage of the swing, about a tenth of a second, the angle between the arm and the shaft is held fixed. This effect can be seen in the diagram of the basic swing in Figure 2.3. The constancy of the angle between the arm and the shaft implies that at this early time there is a twisting force from the wrists on the shaft and the resulting torque prevents the closing of the arm-shaft angle as shown in Figure 2.17.

Fig. 2.16. Showing (a) an angle of 60° between the arms and the shaft early in the swing and (b) how this would lead to a jackknife effect if the wrists were not locked.

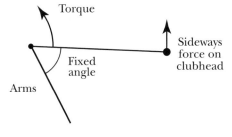

Fig. 2.17. To maintain the angle between the arms and the shaft requires the wrists to apply a torque.

At the start of our chosen swing, the sideways force on the clubhead, perpendicular to the shaft, is about 2 pounds. During the early part of the swing this sideways force decreases and, as described earlier, later in the swing the direction of the force on the clubhead is increasingly aligned with the shaft.

Bending of the shaft

During the swing the force applied to the clubhead results in a bending of the shaft. The largest bend occurs in the early part of the swing when the club is forced forward and the shaft is bent backward as shown in Figure 2.18. A typical backward bend would be about 3 inches.

Fig. 2.18. As the club is forced forward the shaft is bent backwards.

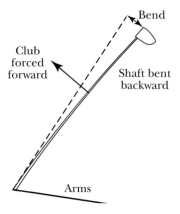

The bending of the shaft can be studied experimentally by fixing the grip end firmly in a vice and loading the shaft at the clubhead as illustrated in Figure 2.19. The amount of bending depends, of course, on the material and construction of the shaft, but a typical shaft would bend about 1.5 inches for each 1-pound load. We see, therefore, that the 3-inch bend of the shaft implies a force of 2 pounds perpendicular to the shaft.

Fig. 2.19. Experimental measurement of the flexibility of the shaft by holding the grip in a vice and applying a load to the clubhead.

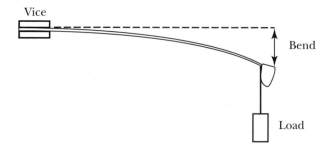

A related experiment measures the natural frequency of vibration of the club. With the grip end of the shaft again held in a vice, the clubhead is displaced and released. The clubhead and shaft then vibrate together and the frequency of the vibration is easily measured. Typical frequencies lie between 4 and 5 oscillations per second, and the difference in frequency between regular and stiff shafts is quite small, with a range of about 15%

Figure 2.20 gives a graph of the bend of the shaft during the forward swing for a typical case. At about half-way into the swing, the torque applied to the shaft diminishes and the shaft begins to straighten. In the final phase the bend reverses and the clubhead moves ahead of the shaft. This is sometimes attributed entirely to the shaft 'springing forward' in response to the earlier backward bend. However, this assumes that the shaft behaves as it would in a vice whereas, at this stage, the hands act more as a pivot.

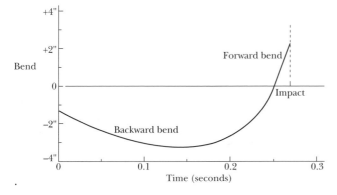

Fig. 2.20. Graph of the bend of the shaft, in inches, during a typical swing, showing the forward bend at impact with the ball.

A substantial contribution to the forward bend arises from the offset of the centre of gravity of the clubhead from the line of the shaft, as illustrated in Figure 2.21(a). This offset is typically about an inch. As we saw earlier, the clubhead is subject to a very large

(a)

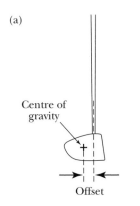

Centre of
gravity

Offset

(b)

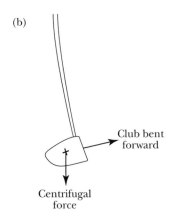

Club bent
forward

Centrifugal
force

Fig. 2.21. (a) The centre of gravity of
the clubhead is offset from the line of
the shaft. (b) The centrifugal force
bends the clubhead forward.

centrifugal force and this force, acting at the centre of
gravity of the clubhead, produces a twisting force that
bends the shaft forward as shown in Figure 2.21(b).

The forward bend alters the effective loft of the club.
The change in the loft angle is about 1.5° for each inch
of bend, so a 2-inch forward bend at impact with the
ball would produce a 3° increase in the effective loft of
the club.

The effect of shaft flexing

At first sight it seems to be an advantage that at impact the clubhead is being sprung forward. However,
the situation is quite complicated. We can illustrate the
problem by a thought experiment in which we assume
that the player is able to deliver a certain amount of
energy to the club during the swing. The energy stored
in the bending of the shaft would then imply a reduced kinetic energy in the clubhead and an associated
reduction in the clubhead speed. In the actual swing,
the effect will be determined by the reaction of the
individual player to the flexing of the shaft, and
the outcome is therefore uncertain.

A more significant factor is the change in the effective loft of the club. If the shaft is bent through a given
angle the effective loft is increased by that amount.
For example the effective loft of a club with a 9° loft
could be raised by 4° to 13°. Whether such changes in
the loft are advantageous depends on the difference
between the loft of the club used by the golfer and his
optimum loft. Accordingly, the range could be increased
or decreased by up to about 10 yards by the flexing of
the shaft. Of course, the golfer who takes these things
seriously will have allowed for this effect in his choice
of the loft of his club.

Forces on the body

We have seen that in the early part of the swing the wrists remain locked while resisting the twisting force as the club is accelerated. The sideways force on the clubhead is typically 2 pounds and we can imagine the effect of this by thinking of holding the club horizontally with a 2-pound weight placed at the clubhead. This weight produces an appreciable torque on the wrists, but balancing the torque is well within their capability.

After this first stage, the force on the hands, and through them to the body, is the pull along the shaft. Although the purpose of the swing is to increase the speed of the clubhead by applying an accelerating force, this force comes to be dominated by the centrifugal force. The total force outward on the club, including the force on the shaft, has to be balanced by an equal and opposite force on the body. As the swing progresses, the strength and direction of the force on the body changes.

In analysing the force on the body, we must recall that the swing lies in a plane at an angle to the vertical so, for example, even at the bottom of the swing there is a horizontal force on the body. Taking a typical swing we find that at the top of the swing there is an upward force of about 10 pounds on the body. When the shaft has reached a horizontal position the force on the body is sideways and has reached about 60 pounds. At the bottom of the swing there is a downward force and a horizontal force, both of about 70 pounds, combining to make a total force of 100 pounds. These forces are largely transmitted to the feet, and as a result they feel an apparent increase in the body's weight of about 70 pounds.

Power

The main purpose of the swing is to propel the ball at high speed. This requires a high clubhead speed and, in turn, the power to provide this speed.

The average power supplied to the club during the swing is equal to the energy of the club at impact divided by the time of the swing. This means that a higher clubhead speed requires more power, both because of the higher energy delivered and because of the shorter time of the swing.

Before analysing the power developed during the swing it is perhaps useful to discuss the units involved. The scientific unit of power is the watt, familiar from its use with electrical equipment. However, it is common in English speaking countries to measure mechanical power in terms of horsepower, the relationship being 1 horsepower = 746 watts. The name arose when steam engines were developed. It was clearly useful to know the power of the engines in terms of the more familiar power of horses.

As would be expected, humans are capable of sustaining only a fraction of a horsepower, a top athlete being able to produce a steady power approaching half a horsepower.

Muscular energy is derived from the breakdown of carbohydrate food stores utilizing atmospheric oxygen. The carbohydrate is stored in the muscle as glycogen and the oxygen is brought to the muscle by circulating blood. The energy so produced is used to make adenosine triphosphate (ATP), which is able to pass on the energy to the muscle.

The muscle's long-term energy requirement needs a continuous supply of oxygen to sustain the complex biochemical reactions that lead to the production of

ATP. The use of the muscles is then limited by the ability of the lungs and the circulatory system to provide
the required oxygen.

However, when short-term power is required, it can
be achieved for a few seconds using the local store of
ATP without calling on the oxygen supply. The forward
swing of the golf club, which takes only a fraction of a
second, clearly relies on this process. Let us now ask
how much power is developed during the swing.

Although the intention of the swing is to deliver
power to the clubhead, this inevitably involves supplying power to the other components of the swing,
the shaft and the arms. The provision of their kinetic
energy requires a significant additional power. The
level of power supplied to each component is obtained
by determining the rate of change of its kinetic energy.
We shall calculate this using measurements from the
photographic study of the swing described earlier.

Figure 2.22 shows the development of the kinetic
energy of the club—clubhead and shaft—during the
swing. The energy is given in joules, a joule being the
energy produced by a power of 1 watt in 1 second.

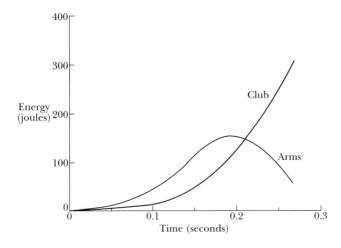

Fig. 2.22. Graph showing the development of the kinetic energy of the
club and the arms during the swing.

It is seen that the energy grows with increasing rapidity, implying an increasing power during the swing. The kinetic energy of the arms is also shown and it is seen that during the first two-thirds of the swing the energy supplied to the arms is greater than that supplied to the club. Thereafter, the arms are slowed and the energy in the arms decreases.

With these graphs it is now possible to calculate the power involved. Figure 2.23 shows the time development of the power supplied to the club together with a graph of the total power supplied to both the club and the arms. The power supplied to the club increases continuously throughout the swing reaching just over 4 horsepower. The total power supplied to the club and the arms reaches a maximum about half-way through the swing. After this, much of the increase in the power supplied to the club comes from the transfer of energy from the arms as the club is pulled forward and the arms are slowed.

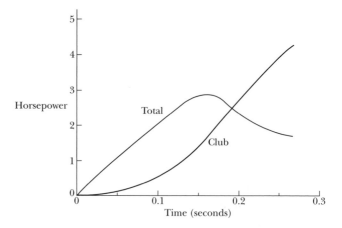

Fig. 2.23. Graphs of the power supplied to the club and the total power supplied to the arms and the club during the swing.

Source of the power

The complicated body movements in the swing combined with the multiplicity of muscles involved makes
it difficult to account for the origins of the power which
is developed. To obtain a rough indication an experiment was carried out to make an estimate of the role of
the muscles of the various parts of the body.

It was arranged that a player would make swings with
a driver with limited body movement. First he made
swings using only the wrists, second he also used his
arms, third he was allowed to move only his body above
the waist, and finally he made a full swing now including the use of his leg muscles. For each set of swings
the clubhead speed at impact was measured and used
to calculate the energy supplied to the clubhead. The
results are given in Table 2.1 where the energy of the
clubhead is given as a fraction of that achieved with a
full swing.

Energy delivered to the clubhead

Wrist alone	10%
Wrists and arms	20%
Wrists, arms and upper body	40%
Complete movement	100%

Although these measurements do not allow a precise
breakdown of the energy supplied during a full swing,
they indicate that the wrists, arms, and upper body
together provide less than half the energy transmitted
to the clubhead, implying that the legs supply more
than half the energy. This outcome might seem surprising because most players are more conscious of the
contribution of the arms than of the legs. However,
the results of the experiment give confirmation of Jack

Nicklaus's comment that 'you hit the golf ball with your legs'.

Effect of gravity

As the club is forced through the forward swing it is simultaneously falling due to the force of gravity. This means that part of the energy received by the club comes from this source. In falling from the top of the swing down to impact, the club's gain in energy is approximately 4 joules. Since the club's total kinetic energy reaches around 300 joules, it is clear that the effect of gravity on the club is very small.

However, the arms also receive energy due to the force of gravity. The arms fall through a smaller distance than the club but they are much heavier. For a typical player the energy supplied to the arms by gravity is around 30 joules. If we compare this with the kinetic energy of the arms as plotted in Figure 2.22, we see that the gravitational force has supplied a significant fraction of the energy of the arms. Some of this energy received by the arms will be transferred to the club but it is difficult to estimate the size of the contribution. However, the energy transferred will be only a very small fraction of the total of 300 joules supplied to the club.

Effect of air-drag

Whereas the force of gravity adds energy to the club during the downswing, the air-drag on the club, of course, extracts energy. It is sometimes conjectured that for the large-headed drivers that have been introduced the increased air-drag could be a significant disadvantage.

There will be a discussion of air-drag in the context of the flight of the ball in Chapter 4. It will be seen there that the air-drag on an object depends on the flow pattern around it. The precise form of the airflow over the clubhead is not known but it is, nevertheless, possible to make an estimate of the magnitude of the air-drag. It turns out that, depending on the form of the airflow, the energy lost by a large clubhead is around 10–20 joules. Rather surprisingly the energy lost through the air-drag on the shaft is comparable to that lost through the clubhead, partly due to the larger area presented by the shaft as compared to the clubhead. A reasonable estimate of the total energy loss through air-drag on the shaft and the clubhead is, say, 30 joules, and this is about 10% of the energy of the club at impact with the ball. This unavoidable loss of energy causes a reduction of roughly 15 yards in the range of the ball.

The drag on the clubhead is proportional to the area of its face. If we take the area of the clubface of a larger club to be 50% more than that of a conventional club, the difference in energy loss is about 5 joules, implying a reduction of a couple of yards in the range of a typical drive.

Dependence of range on the swing

As we shall see in Chapter 7, which deals with the flight of the ball, the range of a drive depends on the angle, speed, and spin with which the ball is launched. With a given club, the variations in the swing mainly affect the speed of the clubhead, and through this determine the launch speed and range of the ball. As mentioned ear-

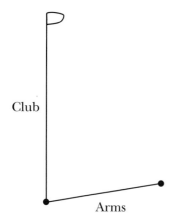

Club

Arms

Fig. 2.24. Example of starting position with reduced swing, the club being held vertically leading to a lower clubhead speed.

lier, the range of a drive is increased by about 3 yards for each increase of 1 mile per hour in the speed of the clubhead at impact with the ball.

In the swing analysed in this chapter, the duration of the swing is 0.27 seconds and the clubhead speed at impact is 107 miles per hour. This swing is a full swing, the backswing being taken back as far as the arms and body allow. It is quite common among golfers that a reduced swing is used, probably to improve the likelihood of a clean hit. This, of course, results in a loss of clubhead speed. We can make an estimate of the speed obtained in the reduced swing by taking the applied force to be the same as the corresponding part of the full swing. For example, if the swing starts with the clubshaft vertical as shown in Figure 2.24, we calculate the subsequent clubhead speed using the same force as that applied at the same positions in the full swing. It turns out that the clubhead speed reaches 94 miles an hour, a reduction from the full swing of 13 miles per hour. This would reduce the range of the shot by about 40 yards

Using this procedure we can calculate the loss of speed for all starting angles of the shaft. Figure 2.25 defines the starting position in terms of the angle of the shaft to the horizontal and gives the loss of the clubhead speed plotted against the starting angle.

Fig. 2.25. Showing the loss of clubhead speed when the angle swung by the clubhead is curtailed.

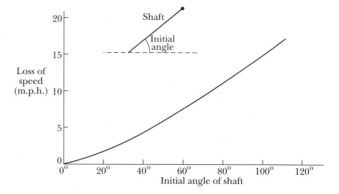

For a full swing, the clubhead speed achieved depends on the force applied by the player. We can see that there is a simple rule here. Since the distance to be travelled by the clubhead is essentially the same for slow and fast swings, the average speed will be inversely proportional to the time taken for the swing. But more than that, if we make the reasonable assumption that for the slower swing the applied force varies in a similar way to that in our basic swing, the clubhead speed at impact will also be inversely proportional to the duration of the swing. For example, if the swing is 10% slower, the duration being, say, 0.30 seconds rather than 0.27 seconds, we can expect the clubhead speed at impact to be reduced from 107 miles per hour to $(0.27/0.30) \times 107 = 96$ miles per hour, a reduction of 11 miles per hour with a reduction in the range of the shot of about 30 yards.

Coordination

We often hear that 'it's all in the timing', but without any convincing explanation of what timing is. The use of the word timing is unfortunate because it implies arranging for something to happen at a particular time. The word is more appropriately used in a game like tennis, where the moving ball must be struck at just the right moment. What is called for in the golf swing is better described as coordination, defined in the *Oxford Dictionary* as 'the harmonious functioning together of different interrelated parts'.

We have seen in this chapter that there are three phases in the swing and what is required is that the transitions between them are smooth and the swing appears as a natural whole. The first phase is the period when the wrists are locked and the angle between the club and the arms remains fixed, the arms and club rotating together as shown in Figure 2.3. During this phase, which lasts for about a tenth of a second, the clubhead speed increases relatively slowly. In the second phase, the club and the arms swing freely at the wrists, and the angle between the arms and the shaft increases. During this phase the centrifugal force comes to dominate and at two-thirds of the time into the swing, the clubhead is being pulled along the line of the shaft. In the third and final phase the clubhead swings outward. The centrifugal force on the clubhead then slows the arms and there is a transfer of kinetic energy from the arms to the club.

When we see a top-class player make a quality swing it appears as a continuous process, but underlying it is the perfect coordination of the movements of the arms, body, and legs to produce a smooth transition through the sequence of events described above.

3 IMPACT

When the clubhead of a driver strikes the ball, the duration of the impact is about half of a thousandth of a second. This is so brief we can learn nothing about the impact from direct observation, and even filming it is quite a challenge, requiring a very high-speed camera. However, it is fortunate that the laws of physics reveal much of what happens during the impact. The first step is to recognize that the process is really a bounce, the ball bouncing off the clubhead. So a good place to start is with the case of a simple bounce of a ball off the ground.

The bounce

In an idealized bounce, the ball is perfectly elastic, the ground is hard and its surface completely smooth and slippery. Figure 3.1 illustrates the bounce for this case.

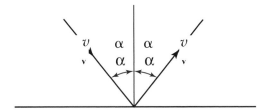

Fig. 3.1. The ideal bounce is symmetric. The speed of the ball is unchanged by the bounce and angle of departure is equal to the angle of incidence.

It is conventional to represent angles by Greek letters and we shall designate the ball's angle of incidence by α (alpha) and its incoming speed by v. When the ball departs it still has a speed v and it leaves at an angle that is equal to the angle of incidence. The first feature of this model that we recognize as unrealistic is that the speed of the ball is unchanged by the bounce. It is a common experience that a ball dropped from a given height does not return to that height after the bounce, and with successive bounces there is a decrease in height until the ball comes to rest.

What really happens is that the ball is deformed during the bounce and some of the ball's energy is dissipated by the internal friction of the ball, the lost energy finally appearing as heat. Another feature, which again is easily observed, is that a ball coming to the ground at an angle with no spin will leave the ground spinning. The spin is generated by friction between the ball and the ground during impact, and the friction force on the ball also produces a third effect—a reduction in ball's horizontal speed. These effects of a real bounce are illustrated in Figure 3.2. The ball's departure speed, v_2, will always be less than its incoming speed, v_1, but the angle of departure, α_2, can be greater or less than the angle of incidence, α_1, depending on the type of ball, the roughness of the surface, and the angle of incidence.

Fig. 3.2. In a real bounce the ball is slowed and spun, and the angle of departure is not equal to the angle of incidence.

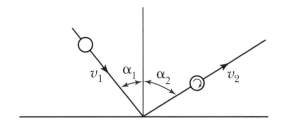

Before using these features of the bounce to analyse the impact of a golf club on the ball, it is of interest to look more closely at the bounce to see what actually happens *during* the brief impact. We shall stay with the bounce from the ground because that is easier to picture. When the ball reaches the ground at an angle it starts to flatten against the ground producing an upward reaction force on the ball, and the ball also slides along the ground as shown in Figure 3.3. The sliding motion results in a horizontal friction force as shown. This slows the ball and also starts the ball rotating. As the deformation increases, the upward force on the ball increases until its vertical velocity is brought to zero. The force then accelerates the ball upward, the area of contact with the ground decreasing until contact is finally lost and the ball leaves the ground.

Fig. 3.3. The ball slides, producing a horizontal friction force, and is compressed, producing an upward reaction force.

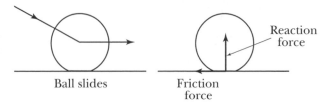

Spin

Regarding spin, there are two cases. If the ball approaches the ground at a sufficiently low angle the ball slides on the ground throughout the impact, the rotation rate increasing during this time. For bounces with a higher angle of approach the rotation rate increases to the value at which there is no relative motion between the surface of the ball in contact with the ground and the ground itself. The ball is then rolling. In this state there is no sliding friction force and the ball rolls throughout the rest of the bounce. It then leaves the ground with the spin rate it had when rolling, as shown in Figure 3.4. We shall find that in the case of the impact of the club on the ball, the ball is normally rolling when it leaves the clubface.

Fig. 3.4. For bounces at higher angles to the ground the friction force stops the sliding. The ball then rolls and finally leaves the ground with the corresponding spin.

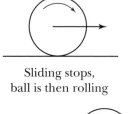

Sliding stops,
ball is then rolling

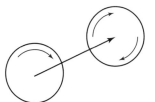

Ball leaves the ground with
"rolling" spin rate

Bounce off the golf club

Turning now to the impact of the club on the ball we first consider the idealized impact with a perfect bounce. The face of the club is at an angle to the shaft, this angle being called the loft. We shall call the loft angle θ (theta). Taking the clubhead to be moving horizontally at impact, and neglecting any bending of the shaft, the ball meets the clubface at an angle θ to the normal as shown in Figure 3.5.

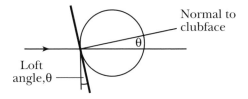

Fig. 3.5. The ball meets the clubface at an angle θ to the normal, where θ is the loft of the club.

However, we have to take care with the calculation of the bounce angle of the ball. It might appear at first sight that, in this idealized case, the ball will bounce off the clubface at an angle 2θ to the horizontal, but this is not so. We start the calculation by looking at the impact from the club's point of view, technically called 'moving to the frame' of the clubhead. In this frame the club is stationary and the ball is seen to be moving toward the clubface as shown in Figure 3.6(a). The resulting bounce is illustrated in Figure 3.6(b).

(a)   (b)

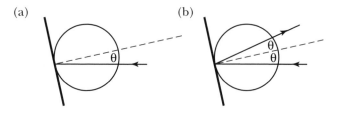

Fig. 3.6. (a) In the frame of the clubhead the ball moves horizontally towards the clubface, at an angle θ to the normal and (b) the ball then bounces off the clubface at an angle 2θ to the horizontal.

The bounce is symmetrical about the normal and it is seen that, in this moving frame, the ball does indeed leave at an angle 2θ to the horizontal. But we want to find the angle that we see in the stationary frame. What we have to do is add the velocity of the clubhead to the velocity the ball has in the clubhead frame. This is done by representing the speed and direction by arrows, called vectors. As shown in Figure 3.7, the ball's velocity, as seen by the clubhead, and the clubhead velocity itself form the sides of a parallelogram, the diagonal of which gives the resulting velocity of the ball. It is seen from the symmetry of the diagram that, in this idealized case, the ball leaves the clubhead at an angle θ to the horizontal, not 2θ. It is also seen from the diagram that the velocity of the ball is greater than that of the clubhead. If we imagine that the ball is struck with a club with zero loft, the clubhead having a velocity v_c, then in the frame of the clubhead the ball would approach the club with a velocity $-v_c$ and leave it with a velocity $+v_c$. This means that, in this idealized case, the ball's velocity has changed by $2v_c$. So an observer would see the ball stationary before impact and having a velocity twice that of the clubhead after impact.

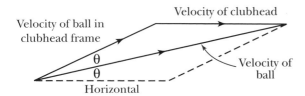

Velocity of ball in clubhead frame

Velocity of clubhead

Velocity of ball

θ

θ

Horizontal

Fig. 3.7. Vector addition of the velocity of the ball with respect to the club and the velocity of the club gives the velocity and angle of the ball's departure. In this idealized case the departure angle is equal to the loft of the club, θ.

The above account of the impact between clubhead and ball is idealized for simplicity, and we shall now consider the modifications needed to have a realistic description.

Coefficient of restitution

We have already noted that a ball loses energy during a bounce. This effect is described quantitatively by introducing the coefficient of restitution. If a ball hits a surface at a right angle with a speed v and leaves the surface with a speed v' then the coefficient of restitution is defined by

$$e = \frac{v'}{v}.$$

So an ideal bounce would have $e = 1$. An example of the opposite extreme is a ball of putty, which dropped to the ground would stay there, having $e = 0$. When bounced off a hard surface, a typical golf ball has a coefficient of restitution of around 0.7. Even for a particular ball there will not be one exact value of e because the value will vary to some extent with the speed of the ball, a slow ball undergoing a smaller deformation than a fast one.

The situation is further complicated when the club-head is not rigid. In this case the flexing of the club-head alters the dynamics of the interaction with the ball. Although this effect complicates the process of the bounce we can still define an effective coefficient of restitution as the ratio of the speed leaving the surface and the speed of approach. However, it is clear that the coefficient of restitution will now be a characteristic of the club–ball combination.

We shall later examine the speed and spin given to the ball on impact with the club but the basic effect of the coefficient of restitution is to reduce the speed given to the ball. If we again imagine a club with zero loft having a clubhead speed v_c, the speed of the ball in the frame of the club will then change from $-v_c$ to $+ev_c$, and the change in the ball's speed is $(1+e)v_c$. So the ball leaves the club with a speed $(1+e)v_c$ as compared with $2v_c$ in the case of a perfect bounce.

The effect of clubhead mass

In the above description of the clubhead–ball impact, it was assumed that the clubhead was so much heavier than the ball that its speed was not changed during the impact. This is a fairly good approximation but there will, of course, be some slowing of the clubhead, with a resulting effect on the speed given to the ball.

For the small loft associated with a driver, the speed given to the ball is essentially given by the equation

$$\text{ball speed} = (1+e) \, \frac{M}{M+m} \times \text{clubhead speed,}$$

where M is the mass of the clubhead, m is the mass of the ball, and the factor $(1 + e)$ allows for the imperfect bounce. Our idealized calculation corresponded to taking the mass of the ball to be negligible compared to that of the clubhead, and in that case

$$\frac{M}{M+m} \rightarrow \frac{M}{M} = 1.$$

A typical value of clubhead mass is 7 ounces (200 grams) and the mass of the ball is 1.62 ounces (46 grams). With these masses,

$$\frac{M}{M+m} = 0.81.$$

So, with a typical driver, allowance for the mass of the clubhead reduces the calculated speed of the ball by about 20%. It is seen that the ball speed for a given clubhead speed could be increased with a heavier club, but it is clear that there is a price to be paid for increasing the clubhead mass, namely that the clubhead speed that can be achieved in the swing will be reduced.

Experiments were carried out by Daish[1] to find how the clubhead speed obtained by players depends on the mass of the clubhead. Using his results, Figure 3.8 gives a graph of clubhead speed against the mass of the clubhead for a player who obtains a clubhead speed of 100 miles per hour with a conventional clubhead.

---

[1] C.B. Daish, *The Physics of Ball Games* (Hodder and Stoughton, 1981).

The figure also shows how the quantity $M/(M+m)$ varies with M. From the equation given above it is seen that the ball speed is proportional to the product of these two factors.

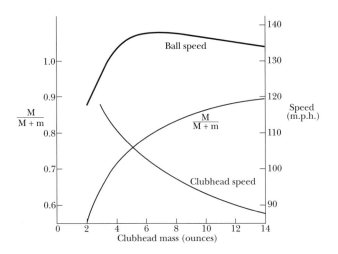

Fig. 3.8. Graph showing the dependence of the clubhead speed and $M/(M+m)$ on the clubhead mass. The speed of the ball is proportional to the product of these two factors.

The ball speed calculated using this equation is also plotted in Figure 3.8, taking $e = 0.7$. It is seen that the maximum ball speed is determined by a compromise between the two factors and is given by a clubhead mass close to 7 ounces. As stated earlier, this is a typical value for the mass of driver clubheads used in practice, presumably arrived at through years of experience.

Although we have examined the case of a player producing a particular clubhead speed with a conventional driver, this was only for illustration. Daish found a similar dependence of clubhead velocity on clubhead mass for a number of players, and the maximization of the ball speed with respect to clubhead mass would give almost the same clubhead mass for all players.

Spin from the clubhead

When the clubhead first makes contact with the ball, the loft of the club causes the ball to slide on the club-face. There is then friction between the ball and the clubface and as the ball becomes more compressed the friction force increases. This slows the face of the ball in contact with the surface of the club and the ball starts to rotate. These stages are illustrated in Figure 3.9. The rate at which the sliding slows and the rotation increases depends on the loft of the club. The critical loft below which sliding stops before the ball leaves the clubface can be calculated, and turns out to be about 75°. Since all clubs have a loft less than this we can assume that the ball is rolling as it leaves the clubface.

The amount of spin produced depends on the loft angle of the clubface and on clubhead speed at impact. The effective loft of the club includes the effect of the bending of the shaft and the combined loft including the effect of bending is called the dynamic loft. Figure 3.10 shows the calculated development of the spin rate during a typical impact for a driver with a dynamic loft of 13° and for a 9-iron with a loft of 46°.

Fig. 3.9. Showing the development of spin as the ball initially slides and then rolls on the club face.

Initially
ball slides

Sliding reduces
and
ball rotates

Sliding stops
and
ball rolls

Fig. 3.10. Graphs of the spin rate during the ball's impact with the club for a driver and a 9-iron.

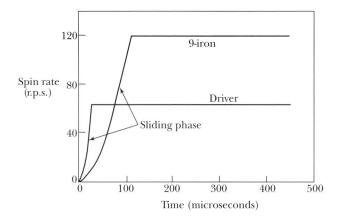

It is seen that the sliding phase is shorter than the rolling phase in both cases. The driver produces a spin of 60 revolutions per second and the 9-iron gives twice that spin. The reduction of the spin rate during the flight of the ball is quite small, and in a drive with a flight time of 8 seconds the ball will rotate about 400 times.

It is also of interest to ask how far the ball moves on the club during impact. This turns out to be about an eighth of an inch for the driver and a quarter of an inch for the 9-iron. In both cases the movement is too small to have a significant effect on the shot.

We recall that for an idealized bounce from the clubhead, with the assumption that mass of the ball is negligible compared to that of the club, the ball would leave the club along the normal to the clubface. When we take the realistic case, allowing for the coefficient of restitution, for the proper masses of the clubhead and ball, and for the production of spin, the launch angle is slightly lowered, and for a ball struck horizontally the launch angle is about 0.8 of the dynamic loft as illustrated in Figure 3.11.

Fig. 3.11. In the realistic case the launch angle is reduced to about 0.8 of the dynamic loft angle.

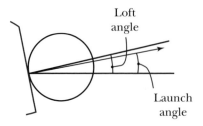

Inside the ball

Up to this point the simplified discussion of the impact of the club on the ball has perhaps given the impression that the ball remains essentially spherical and that the mechanics of the impact involve only the surface of the ball. Nothing could be further from the truth, at least in the case of the drive and other hard hits. In such shots the ball is substantially deformed during the impact as is made clear by the flash X-ray photograph in Figure 3.12.

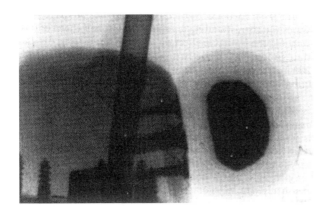

Fig. 3.12. Flash X-ray photograph taken during impact of the club-head with the ball. *Source*: C.B. Daish.

The sequence of events during impact can be examined using the high-speed camera photographs shown in Figure 3.13. The framing rate is 10,000 per second, giving 100 microseconds between frames. The total time of contact between the clubhead and the ball is about half a thousand of a second, which is 500 microseconds.

Fig. 3.13. Frames from high-speed camera photography of the impact of club and ball. *Source*: Photron (Europe) Limited and Department of Sports Science, University of Loughborough.

The behaviour is summarized in Figure 3.14(a). When the clubhead first touches the ball there is no force on the ball, the area of contact being essentially zero. As the clubhead moves forward it starts to compress a layer of the ball immediately in front of the clubhead while most of the ball remains stationary. The 'information' that the ball has been hit travels through the ball in a shock wave travelling at approximately the speed of sound in the material of the ball, and the compression of the ball follows behind this shock. This is seen at 150 microseconds in Figure 3.14, at which time the rear half of the ball is compressed. In most solids the speed of sound is very high, a typical value being 7000 miles per hour. At this speed sound would travel across the 1.68-inch diameter of the ball in 13 microseconds. The material of the golf ball, although hard to the touch, is much softer, with a much lower sound speed. At this slower speed the shock wave travels across the ball in about 250 microseconds, the ball then being fully compressed. The ball then behaves like a compressed spring and starts to expand, as shown at 350 microseconds, finally leaving the clubface at about 500 microseconds.

Figure 3.14(b) gives a graph of the horizontal deformation of the ball during the impact. We see that maximum compression occurs half-way through the contact time, and the deformation of the ball is then a third of an inch. At this time the diameter of the circle of contact between the ball and the clubface is about an inch.

We can now see how the ball acquires a higher speed than the clubhead. At maximum compression the ball and the clubhead are moving together, but the ball has the stored energy of the compression. By the end of the impact this energy has been released, springing the ball away from, and ahead of, the clubface. Because the material of the ball has some 'viscosity', part of the energy used in the compression is lost as heat, and this accounts for the coefficient of restitution being less than one and the ball leaving the clubhead with a correspondingly reduced speed.

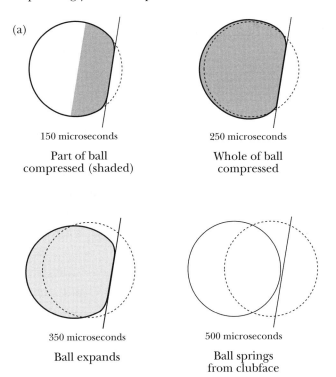

(a)

150 microseconds

Part of ball
compressed (shaded)

250 microseconds

Whole of ball
compressed

350 microseconds

Ball expands

500 microseconds

Ball springs
from clubface

Fig. 3.14. (a) Compression and expansion of ball and (b) a graph of its deformation during impact.

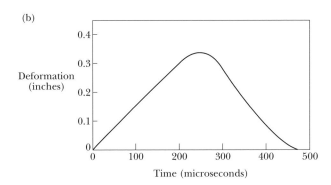

(b)

Deformation
(inches)

Time (microseconds)

Transfer of energy

When the club first makes contact with the ball, all of the available energy is in the energy of motion of the clubhead, its kinetic energy. During the impact, part of this energy is transferred to the ball. This is in the form of the kinetic energy of the ball's forward motion and its spin, together with the internal energy of the ball. The internal energy consists of the energy of the ball's compression together with the energy dissipated as heat. The energy of compression is finally released partly as kinetic energy of the ball and partly as heat. The resulting temperature rise is quite small, less than a degree Celsius.

Although spin is extremely important in the flight of the ball, the kinetic energy associated with the spin is small, less than 1% of the clubhead energy. We can, therefore, neglect the energy of the spin in the energy balance.

Figure 3.15 gives the result of a calculation of the time variation of the energies for the case of a drive. Overall, the clubhead gives up approximately 50% of its energy to the ball, with 40% as the desired kinetic energy of the ball and 10% as heat.

Fig. 3.15. Graphs showing the transfer of energy from the clubhead to the ball during impact.

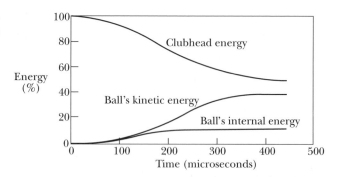

So we can say that the overall efficiency of the hit is about 40%. There is very little that can be done to significantly change this value. It is the result of optimization of the design of the clubhead and the ball over many years, together with the constraints of the laws of mechanics.

Force on the ball

The force on the ball during the brief impact with the clubhead is phenomenal. The force can be calculated using Newton's law in the form,

force = rate of change of momentum.

The momentum given to the ball is mv, the product of its mass m and its velocity v. Since this momentum is given to the ball during the contact time, say half a millisecond, we can obtain the average force by dividing mv by this time. During the impact the force rises from zero to a peak value and then falls back to zero, the peak value being about twice the average. For a ball given a velocity of 120 miles per hour the peak force turns out to be about a ton, which is more than 20,000 times the weight of the ball. Putting it another way, the acceleration of the ball is more than 20,000g where g is the acceleration due to gravity. It is surprising that the ball can suffer this impact with no damage at all.

The effect of spin

The spin given to the ball during the impact plays a very important part in the flight of the ball. The backspin produces a lift which works against gravity and side-spin produces a sideways force which leads to a deflection of the ball. The history of the understanding and explanation of the effect of spin will be left to Chapter 5, which is devoted to this subject.

In the next section we shall look at mis-hits, the predominant effect of which is to give the ball a side-spin. In order to understand the direction of the bend this gives to the flight, we need to know the direction of the force produced by the spin and this is illustrated in Figure 3.16. The effect is similar to the deliberate bending of a shot in soccer, achieved by kicking the ball off-centre.

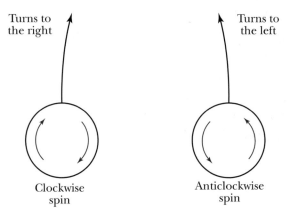

Fig. 3.16. Showing the direction of the bend given to the flight of the ball by clockwise and anticlockwise spins.

Turns to the right

Turns to the left

Clockwise spin

Anticlockwise spin

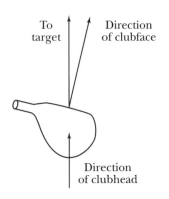

Fig. 3.17. The angled clubhead mis-hit.

To target

Direction of clubface

Direction of clubhead

Mis-hits

We shall first identify the three basic types of mis-hit and then examine them in detail. The simplest mis-hit is where the club is moving in the direction of the target but the clubface is not square to the target, as shown in Figure 3.17. The ball then moves off in the wrong direction, with a spin that produces a curved flight and accentuates the error.

The second type is the directional mis-hit in which the clubface is square to the target but the direction of the motion of the clubhead is at an angle to the target as shown in Figure 3.18. The error in the initial direction of the ball is much less than the error in the direction of the club, mainly because the ball rolls on the club-face during impact. However, this rolling also produces a spin that over-corrects the error, taking the ball the other way and further off-target.

The third type of mis-hit is the off-centre hit. In this case, the clubhead is square to the target and the direction is also correct. The problem arises because the point of contact is not on the line of the centre of gravity of the clubhead, as shown in Figure 3.19. This causes the clubhead to twist, sending the ball away at an angle and with a spin that again produces a curved trajectory. We shall now examine these mis-hits in turn.

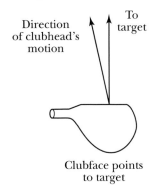

Fig. 3.18. The directional mis-hit – the swing off-line.

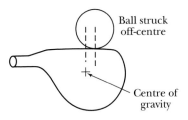

Fig. 3.19. The off-centre mis-hit.

The angled clubhead mis-hit

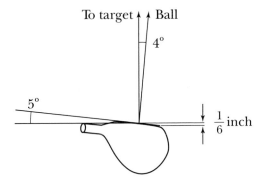

Fig. 3.20. An example of the angled clubhead mis-hit with a 5° tilt.

The angled clubhead mis-hit sends the ball away at an angle to the target which is slightly less than the tilt angle of the club. The case of a 5° clubhead angle is illustrated in Figure 3.20, the ball leaving the clubhead about 4° off-target.

However, the misdirection of the shot is only part of the problem—the angled clubhead also gives the ball a spin which, by creating a sideways force, introduces an additional error. If we take a case where the clubface is angled at 1° in a 200-yard drive, the ball leaves the clubface at an angle which would take it about 3 yards off-line and is given a side-spin of about 4 revolutions per second. At this low spin rate, the force on the ball is uncertain but using a reasonable estimate the curved flight of the ball takes it a further 12 yards off-line, as illustrated in Figure 3.21.

Fig. 3.21. Showing a typical trajectory for an angled clubhead mis-hit with a 1° tilt.

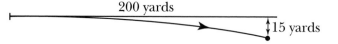

200 yards

15 yards

Clubhead swing off-line

The most usual form of this mis-hit is the dreaded slice. Many beginners find that their drive starts off in roughly the desired direction but veers alarmingly to the right. For some players this problem seems incurable. The slice is caused by the clubhead being pulled across the ball so that, even if the clubface is square to the target, the ball is given a spin.

Skilled golfers can use this shot in a controlled way to produce a desired curvature of the ball's flight. It is then given the more polite name of a fade rather than a slice. Similarly, by directing the club 'from in to out', it is possible to intentionally produce a shot that curls to the left. This is called a draw. For a left-handed player, of course, the directions are reversed.

The seriousness of the slice is brought out in Figure 3.22 which gives the result of a calculation of the amount by which sideways shift of the ball at the end of a 200-yard shot increases with the off-line angle of the clubhead direction. Healthy slices into the wilderness must provide a good income for golf ball manufacturers.

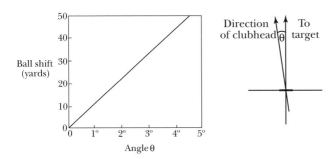

Fig. 3.22. Graph of the sideways shift for a 200-yard shot resulting from a swing off-line by an angle θ.

Off-centre mis-hit

The first two mis-hits we have looked at were quite easy to understand. The mis-hit in which the ball is hit off-centre is surprisingly complicated. Although most golfers are unaware of it, hits off-centre on a flat-faced club would lead to very large errors in the flight of the ball. It is only because designers and manufacturers have produced clubs that provide an antidote that these errors are corrected.

(a)   (b)

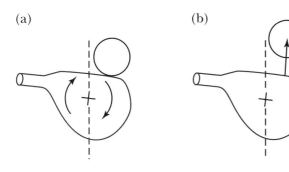

Fig. 3.23. An off-centre hit causing the clubhead to twist and the ball to leave in the wrong direction.

We shall start by examining the effect of hitting the ball off-centre towards the toe of a flat-faced club. A hit towards the neck would have a similar but opposite result. Figure 3.23(a) illustrates the effect the off-centre hit has on the club. The impact with the ball causes the clubhead to twist so that the ball is being hit by a tilted surface. This, of course, causes the ball to leave the club at an angle to the desired direction as shown in Figure 3.23(b). However, the resulting error is quite small, typically a few yards. The main effect comes from the spin given to the ball. This spin produces a sideways force on the ball during its flight, and for a hit that is only half-an-inch off-centre this force would typically lead to a deflection of about 30 yards—and this is for a shot in which the club meets the ball on the correct line with the clubface square to the target.

What is surprising, because it is counter-intuitive, is the direction of the spin and the resulting effect on the ball. It might be thought that in the case illustrated in Figure 3.23 the ball would roll off the tilted clubface with a clockwise spin. In fact the spin is anticlockwise. The reason for this is brought out in Figure 3.24.

Fig. 3.24. In the off-centre mis-hit the twist of the clubhead turns the ball as a gear would do.

As the clubhead tilts, the clubface moves across the ball and in the process turns the ball as shown in the figure.

Because this is reminiscent of the way in which meshed gears turn on each other it is usually called the gear effect.

Let us now look at the trajectory that results from this angling of the shot and the curved flight caused by the spin of the ball. Figure 3.25 shows the calculated trajectory for a shot hit half-an-inch off-centre and subject only to the gear effect. It is seen that after the flight and run of the ball it is almost 30 yards off course.

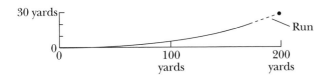

Fig. 3.25. Shows the trajectory that a ball hit with a flat-faced club would take as a result of the ball being hit off-centre by half an inch for a 200-yard drive.

Now, many of us regularly hit the ball off-centre but we do not see this large deviation from the desired direction. The reason is that clubs are made in a way which, in large part, corrects for the effect of the tilting clubhead. The correction is achieved by making the clubface curved, so that looking down on it the face has a small outward bulge. The radius of curvature of the clubface for drivers is typically around 10 inches. Figure 3.26 shows the effects to be expected from this curvature if it operated alone. The clubface strikes the ball at an angle and the effects are similar to those produced by the loft of the club, the ball leaving the club-face at an angle and with spin, in this case side-spin.

We now need to combine the effect of the clubface curvature with that of the tilting causing the mis-hit. The departure angles for the two effects are in the same direction and will add together to work against the large deviation arising from the gear effect. However, the main correction comes from the spin produced by the curvature of the club's surface. If this curvature

Fig. 3.26. Showing the direction of the spin given to the ball by the curvature of the clubface in an off-centre hit.

is just right the deflection due to curvature will balance that due to the tilting, and the overall result will be that the ball will end up in the middle of the fairway. In the example we have taken this is calculated to occur for a clubface with a radius of curvature of 10 inches and the resulting trajectory is shown in Figure 3.27.

Fig. 3.27. Showing how the curvature of the clubface corrects the gear effect of an off-centre mis-hit.

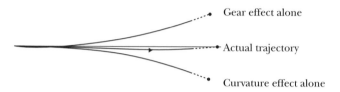

Gross mis-hits

The mis-hits considered above occur at least to a small degree with most shots. However, occasionally the ball is hit so badly that it comes off the edge of the club as illustrated by the examples in Figure 3.28. In the first case the ball is 'skied' and in the second the topped ball runs along the ground.

Fig. 3.28. Gross mis-hits in which the ball is skied and topped.

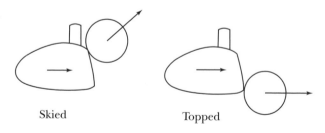

When the ball is struck in a drive it is substantially squashed and the contact between the ball and the clubface is circular. During the impact the diameter of this circle typically grows to about an inch. This means that in some shots this circle will overlap the edge of the club with detrimental effects of varying degree. Figure 3.29 shows a case with such an overlap.

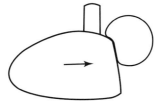

Fig. 3.29. Deformation of the ball in a gross mis-hit.

Figure 3.30(a) shows the area of a conventional driver which is affected by this type of mis-hit. The shaded area shows the region of central points of impact for which shots are unaffected by the edge of the club. For less good players one of the advantages of the now popular larger clubheads is the substantially larger area it allows for hits that are unaffected by overlap of the club edge, as shown in Figure 3.30(b).

(a)   (b)

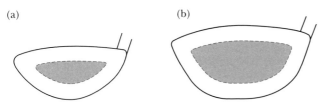

Fig. 3.30. The shaded area shows the region of initial contact on the clubface where the deformation of the ball is unaffected by overlap of the edge of the clubface for (a) a conventional driver and (b) a large-headed driver.

4 AERODYNAMICS –DRAG

The aerodynamics of golf balls is an intriguing subject and, as we shall see, it holds several surprises. The most obvious effect of the air on the flight of the ball arises when there is a strong wind. A side wind can turn a shot aimed at the fairway into a search for the ball in the rough grass. A head wind leads to a disappointing shot, and a strong following wind gives a golfer the opportunity of producing his longest ever drive. However, although these are familiar effects they are minor compared to the less obvious but stronger effects of drag and lift on the ball as it flies through the air, with or without a wind.

The magnitude of the drag can be illustrated with a simple example. Top-class hitters can send the ball away at 160 miles per hour to return to earth 300 yards down the fairway. If we calculate the range of a ball hit at this speed neglecting the effect of the air, we find that a shot at 45° would have a range of 570 yards—and then the bounce and roll. Such a drive would reach the green on almost all par five holes.

Newton's calculation

The physics of air-drag was first studied by Isaac Newton, and he even carried out a calculation of the drag on a sphere. In this calculation he imagined that as the ball moves through the air it collides with stationary air particles. This transfers momentum to the particles and results in a drag on the sphere. This was a clever calculation for the seventeenth century and, although the mechanics envisaged by Newton is not right, his calculation brings out the important factors. He found, correctly, that the drag force is proportional to the density of the air and to the cross-sectional area of the sphere. He also found that the force is proportional to the square of the ball's speed. There is some truth in this, but in reality the variation with speed is much more complex. Anyway, Newton's formula is nowadays taken as basic and the actual behaviour, determined by experiment, is described in terms of the deviation from this formula.

The mistake in Newton's calculation was the assumption that the sphere would have independent collisions with each of the air particles. In fact the air is made up of tiny molecules which are so numerous they are continually colliding with each other. In a sphere the size of a golf ball, there are a thousand, million, million, million (10^{21}) molecules. The molecules are moving in random directions with speeds of a thousand miles per hour, each having more than a thousand, million collisions per second.

It is, of course, impossible to deal with the individual motions of this vast number of particles, and it turns out that the best way of understanding the behaviour of the air is to regard it not as a collection of molecules but as a fluid. As the golf ball moves along its path it has to push this fluid aside, the resulting force on the ball being proportional to the density of the air.

The density of air is measured in mass per unit volume. As mentioned earlier, there is often confusion between mass and weight, but we recall that the weight of a given mass is simply the force of gravity on that mass, a mass of 1 pound being said to have a weight of 1 pound. The air is of course very light, a cubic foot having a mass of 1.2 ounces. However, in a golf drive the 1.68-inch-diameter ball typically sweeps through a total volume of about 10 cubic feet, the air in this volume having a mass of 12 ounces. So we see that the ball has to sweep through a mass of air which is much greater than the mass of the ball, 1.62 ounces, and this produces the substantial drag.

The airflow

In studying the flight of the ball it is much easier to understand the interaction between the ball and the air by looking at it 'from the ball's point of view'. So rather than trying to follow the ball through the air we take the ball to be stationary and the air to be flowing over it. The picture is then the same as that in wind tunnels used to study the airflow over stationary models, air-

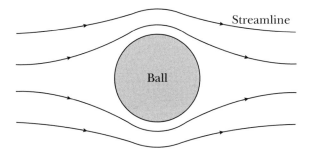

Fig. 4.1. Cross section of the airflow over the ball, the flow following the streamlines.

craft wings for example. Figure 4.1 gives a simplified illustration of the airflow over the ball. The lines of flow are called streamlines. Each element of air on a streamline stays on that streamline and the air between two streamlines remains between those streamlines. The figure actually shows a cross section through the ball. In a three-dimensional view, as shown in Figure 4.2, we see that there are actually stream surfaces, the streamlines lying in these surfaces.

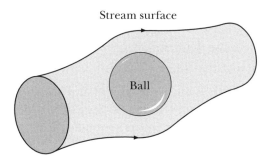

Fig. 4.2. Three-dimensional drawing of a stream surface, showing how the air flows around the ball.

D'Alembert's paradox

If we take the simplest fluid theory of the flow we find that the flow pattern is just like that of Figure 4.1. However, there is a surprise. When the French mathematician d'Alembert studied such flows in the eighteenth century he found that there is no drag. This can be understood from the figure, where we see that the downstream flow is identical to the upstream flow. This means that no momentum has been transferred from the ball to the air and consequently there is no force on the ball. This problem is known as d'Alembert's paradox.

The paradox is resolved by the recognition that the simple fluid theory neglects the viscosity of the air, and that in reality the role of viscosity is fundamental. We are more familiar with viscosity as a property of liquids, such as oil. Air has a much lower viscosity, but it is made apparent, for example, in the slowing of the air which has left a fan.

Stokes's model

A model of viscous flow over a sphere was provided by the Irish physicist Stokes in the nineteenth century. In this model it is assumed, correctly, that the flow velocity at the surface of the sphere is zero. Well away from the sphere the flow is uniform, being unaffected by the presence of the sphere. The flow between these two regions is governed by viscosity. Many physics students have studied the rate of fall of small spheres through oil or glycerine, using Stokes's formula for the drag to determine the viscosity of these liquids.

In Stokes's model the flow of the fluid past the sphere is smooth and the viscous effects of the sphere extend out to distances comparable with the size of the sphere itself. The problem with this model is that it is only valid for small speeds of the sphere. In fact the theory fails for golf ball velocities greater than 4 feet per hour! This is clearly of no practical interest for golf and we have to think again.

So what happens at higher speeds? Stokes's theory predicts that as the speed is increased the region affected by viscosity reduces in size and for the speeds which are of interest, say 100 miles per hour, the viscous effect is restricted to a region less than a millimetre thick at the surface of the ball. It might be thought that such a thin layer would have a negligible effect whereas, in fact, the behaviour of this layer is crucial.

The boundary layer

The whole issue of viscous flow around solid bodies was resolved by the German physicist Prandtl at the beginning of the twentieth century. The thin viscous region around the surface is called the boundary layer. Prandtl explained that the boundary layer does not continue all the way round the sphere, but separates from the surface at the back of the sphere as shown in Figure 4.3. This separation of the flow produces a wake behind the sphere. The air in this wake has become turbulent and in the process has slowed down. It is the reaction of a ball to this slowing of the air that is the source of the drag on the ball. To see how this happens we need to understand how the air speed changes as the air flows around the ball, and how this is related to the variation in the pressure of the air. This leads us to the theory of Bernoulli, a Swiss mathematician, which deals with this relationship between speed and pressure.

Ludwig Prandtl (1875–1953).

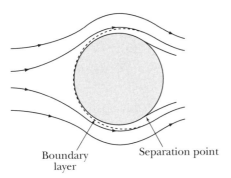

Boundary layer

Separation point

Fig. 4.3. The boundary layer is a narrow region around the surface of the ball in which the effect of viscosity is concentrated. Viscosity slows the airflow causing it to separate from the ball.

The Bernoulli effect

Figure 4.4 shows streamlines for an idealized flow. If we look at the streamlines around the ball we see that they crowd together as the air flows round the side of the ball. For the air to flow through the reduced width of the flow channel it has to move faster. The air speeds up as it approaches the side of the ball and then slows again as it departs at the rear.

Fig. 4.4. To maintain the flow where the channel between the streamlines narrows at the side of the ball, the air has to speed-up. It slows again as the channel widens behind the ball. Pressure differences arise along the flow to bring about the necessary acceleration and deceleration.

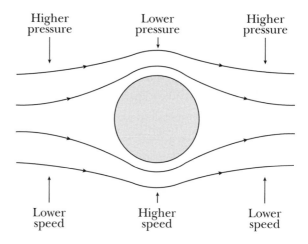

For the air to be accelerated to the higher speed, a pressure difference arises. The pressure in front of the ball is higher than that at the side, and the pressure drop accelerates the air. Similarly a pressure increase arises at the back of the ball to slow the air down again. This effect can be seen more simply in an experiment where air is passed through a tube with a constriction as illustrated in Figure 4.5. For the air to pass through the constriction it must speed up and this requires a pressure difference to accelerate the air. Consequently, the pressure is higher in front of the constriction. Similarly, the slowing of the air when it leaves the constriction is brought about by the higher pressure downstream. If pressure gauges are connected to the tube to measure the pressure differences, they show a lower pressure at the constriction where the flow speed is higher. It is this relationship between speed and pressure that was formulated precisely by Bernoulli.

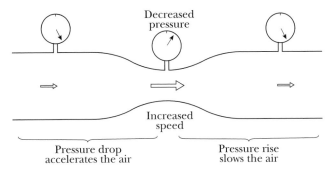

Decreased
pressure

Increased
speed

Pressure drop
accelerates the air

Pressure rise
slows the air

Fig. 4.5. In this experiment air is passed down a tube with a constriction, and pressure gauges measure the pressure changes. The pressure falls as the flow speed increases, following Bernoulli's law.

Separation of the flow

Returning now to the real situation with separation of the flow from the surface of the ball, we ask why this separation happens. As we have seen, pressure differences arise to accelerate and decelerate the flow, but in the boundary layer viscosity also slows the air. This removes the front to back symmetry of the flow. The air is halted before reaching the rear of the ball, and the flow separates from the surface.

This effect has been compared with that of a cyclist free-wheeling down a hill into a valley. His speed increases until he reaches the valley bottom. If he continues to free-wheel up the other side, the kinetic energy gained from going down the hill is gradually lost and he finally comes to rest. If there were no friction he would reach the same height as his starting point, but with friction he stops short of this.

Similarly, the air in the boundary layer accelerates through the pressure drop and decelerates through the pressure rise. Viscosity introduces an imbalance between these parts of the flow and the air fails to complete its journey to the back of the ball. Figure 4.6 shows how the forward motion of the air is slowed and the flow turns to form eddies.

Fig. 4.6. Viscosity slows the separated airflow, producing eddies behind the ball.

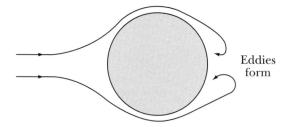

The turbulent wake

The flow beyond the separation is irregular. Figure 4.7 illustrates the turbulent eddies which are formed, these eddies being confined to a wake behind the ball. The eddies in the flow have kinetic energy, and this energy has come from the ball's loss of energy as it is slowed by the air-drag. With increasing ball speed the drag force initially increases as the square of the speed, doubling the speed producing four times the drag. However, with further increase in speed there is a surprising change, and above a certain critical speed the drag force behaves quite differently.

Fig. 4.7. The separated flow is unstable and forms a turbulent wake.

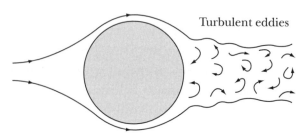

The critical speed

There have been precise experimental measurements of the drag on smooth spheres. This allows us to calculate the drag on a smooth sphere the size of a golf ball and to determine how it varies with speed. The result is shown in Figure 4.8. It is seen that there is an abrupt change at around 250 miles per hour. Immediately above this critical speed the drag force actually falls with increasing speed, dropping to about a third of its previous value at a speed of just over 300 miles per hour before increasing again.

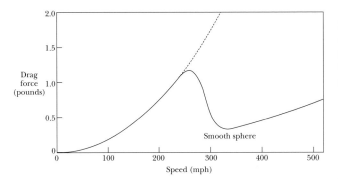

Fig. 4.8. The graph shows how the drag force varies with speed for a smooth sphere the same size as a golf ball. The dashed line gives a (speed)^2 extrapolation.

The speed of golf balls is much lower than this critical speed and we would, therefore, expect that the associated large drop in drag is irrelevant for golf. However, there is a remarkable twist in the story, which makes the behaviour at the critical speed a crucial factor in the flight of golf balls.

The golfers' discovery

The discovery that the critical speed is important for golf balls has its origins in the nineteenth-century game. In earlier times golf balls had been made by

Guttie (c. 1860)

Hammered Guttie (c. 1870–1880)

Bramble Pattern (1890)

Early Dimple pattern

Modern Dimple Pattern

(Courtesy of the Acushnet Company)

stuffing feathers into a leather casing, but around 1850 the 'guttie' ball appeared. This, much cheaper, ball was made from a natural gum called gutta-percha. The ball was hard with a smooth surface. Unfortunately the guttie did not fly as far as the 'featherie'. But surprisingly, as the ball became worn and the surface rougher, the ball would travel further. The natural response was to hammer a roughness onto the surface of new gutties rather than wait patiently for the roughness to appear in play.

It soon became clear that it was easier to incorporate the "roughness" of the surface in the manufacturing process. With one popular ball, small bumps were moulded on the surface giving the impression of a bramble. Then, in 1908, the English engineer William Taylor took out a patent on an 'inverted bramble' pattern having a large number of dimples spread over the surface. Although variations in the dimple pattern have been tried over the years the modern ball has essentially the same character as Taylor's invention.

These advances in ball design were all made on an empirical basis, the underlying physics being a complete mystery. We now know that the explanation involves the critical speed described earlier. As we shall see, the roughening of the surface leads to a substantial lowering of the critical speed. For the modern dimpled ball the critical speed is about 30 miles per hour, and this means that the reduction in drag above the critical speed is made available to the golf ball. The behaviour is particularly surprising since intuition suggests that roughness of the surface would increase, not decrease, the drag.

Figure 4.9 gives the dependence of drag on speed for a golf ball and compares it to the drag on a smooth sphere of the same size. It is seen that for all speeds of interest the dimples reduce the drag to about a half. For example, the drag on a golf ball travelling at 100 miles per hour is reduced from a fifth of a pound to a tenth of a pound. That this is a very important reduction is clear from the fact that even the reduced drag force at 100 miles per hour is approximately equal to the weight of the ball since the weight of the ball is close to 1.6 ounces, which is a tenth of a pound. We see that a golf ball travelling at more than 100 miles per hour sees a drag force which is larger than its weight.

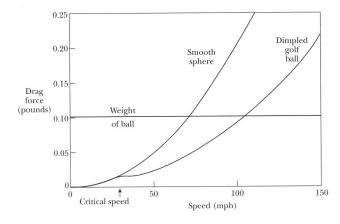

Fig. 4.9. Drag force on the golf ball. Above the critical speed the drag falls below that for a smooth sphere.

We now have two questions to examine. First, what is the cause of the fall in the drag at the critical speed and, second, why is the critical speed lowered by the roughening of the surface?

What happens at the critical speed?

The change in drag above the critical speed arises from a change in the pattern of the airflow. Above the critical speed the narrow boundary layer at the surface of a smooth sphere becomes unstable as illustrated in

Figure 4.10. This allows the faster moving air outside the boundary layer to have a turbulent mixing with the slower air near the surface of the ball and to carry it further towards the back of the ball before separation occurs. The result is a smaller wake and a reduced drag.

Fig. 4.10. Above the critical speed the boundary layer becomes turbulent and this delays separation, reducing the wake and the drag.

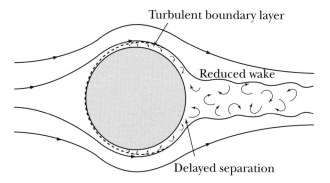

The onset of the instability and the turbulent mixing depends on the roughness of the surface of the sphere. The dimples covering the surface of the golf ball perturb the smooth flow of the air, leading to the earlier onset of turbulence in the boundary layer and a lowering of the critical speed. There are typically about 400 dimples on a golf ball. The dimples are quite shallow, having a depth of about a hundredth of an inch, allowing them to affect the boundary layer, which is of comparable thickness.

The fortuitous discovery of the reduced drag has transformed the nature of golf. Without the dimples the range of a long drive would be reduced by around 100 yards. However, although the effect of drag on the flight of the ball is crucial, it is only half of the story. The lift arising from the interaction of the spin of the ball with the airflow is equally important and we shall turn to this in the next chapter. But first we must complete the account of drag by looking at the effect of atmospheric conditions.

The effect of atmospheric conditions

The influence of the atmosphere on the flight of the ball depends on its density. There are three conditions of the atmosphere that affect its density, namely the atmospheric pressure, the temperature, and the humidity. Dry air at 70°F and normal atmospheric pressure has a density of 2.0 pounds per cubic yard. The density increases in proportion to the atmospheric pressure and decreases in inverse proportion to the absolute temperature, which is measured from the absolute zero of temperature, −460°F.

In Britain the atmospheric pressure typically varies by ±3% about its average value and the absolute temperature, over a reasonable range of playing temperatures, has a variation of about ±5%. There is, however, a correlation between atmospheric pressure and temperature, and the resulting variation in density is roughly ±4% about the mean. The corresponding variation in drag is, therefore, also ±4%.

The effect of humidity is an interesting subject. Sports commentators often suggest that more humid air is denser because of the extra water vapour it holds. The usual response of those who have thought about it is that this is wrong. The argument is that water molecules (H_2O) are lighter than nitrogen (N_2) and oxygen (O_2) molecules, having a molecular weight of 18 as compared to 28 and 32, and so humid air will be lighter and not heavier than drier air. Assuming equal pressures and temperatures this argument is correct. However, the facts are more complicated.

We certainly find that the atmosphere is usually lighter on days of higher humidity, but the change is many times greater than that predicted for the substitution of water molecules for air molecules. The reason is that higher water vapour content tends to occur on days of higher temperature, and higher air temperature is associated with lower air density. The resulting changes in atmospheric density are much greater than the changes due to humidity. So, although the density variations of the atmosphere are correlated with humidity changes, they are not caused by them.

However, although it is interesting to understand these density variations, they are not really very significant. A 3% density change would typically alter the range of a drive by about 2 yards and it seems doubtful whether a golfer would detect this small difference given the general variation of range he experiences.

Another effect on the drag arises when the game is played at higher altitude. The density of the atmosphere decreases with height, initially reducing by 1% for each 340 feet of altitude. On high altitude golf courses the reductions in drag and lift are substantial, and the resulting effect on the range of drives will be discussed in Chapter 7.

Finally, we can ask about the effect of rain. When it is raining the ball will collide with the raindrops and in the process will be slowed. The ball is generally moving much faster than the raindrops and so the raindrops can be regarded as essentially stationary. When the air is pushed aside to go around the ball it produces a viscous drag on the raindrops, pulling them in the direction of the airflow. However, the viscous force is too weak for the raindrops to be carried with the air, and the raindrops in the ball's path collide with the ball.

It is rather nice that Newton's calculation of the drag is applicable in this situation, with the raindrops now replacing his air particles. It turns out that typically the ball will hit just a few rain drops during its flight, and each raindrop has a mass of only about one-ten thousandth that of the ball. The resulting drag and slowing of the ball by raindrops is therefore very small and the reduction in the range of a drive is generally less than a yard. It seems likely that the effect of rain on a player's ability to make his shot will be much more important.

5 AERODYNAMICS —LIFT

When the ball leaves the clubhead it has a high rate of spin. A typical drive gives the ball a back-spin of around 60 revolutions per second, and the higher loft of an iron can produce a spin rate more than twice as fast. The principal effect of this spin is to produce a force on the ball perpendicular both to the axis of the spin and to the direction of the ball. The combination of the back-spin given to the ball in a drive and the ball's forward velocity produces an upward force on the ball which can be greater than the downward force of gravity. In this case the curvature of the initial flight of the ball is upwards rather than downwards.

The phenomenon is now quite well understood, but it took time to discover the proper scientific explanation. The stages in the development of our understanding make an interesting story and we shall briefly look back at the key events.

As with many subjects Isaac Newton was the first to make a contribution. In a letter to the Royal Society in 1672 he describes his theory of light and at one point uses an analogy in which he mentions

'I remembered that I had often seen a Tennis ball, struck with an oblique Racket, describe such a curve line. For, a circular as well as a progressive motion being communicated to it by that stroak, its parts on that side, where the motions conspire, must press and beat the contiguous Air more violently than on the other, and there excite a reluctancy and reaction of the Air proportionably greater.'

The matter was left there until the next century, when the English mathematician and engineer Benjamin Robins studied the flight of musket balls. In 1742 he described his observations and gave his conclusions in a book entitled *New Principles of Gunnery*. He reported that

'the same Piece, which will carry its Bullet within an Inch of the intended Mark, at 10 Yards Distance, cannot be relied on to 10 Inches in 100 Yards...Now this inequality can only arise from the Track of the Bullet being incurvated sideways...What can be the Cause of a Motion so different from what has been hitherto supposed?... its general Cause is doubtless a whirling Motion acquired by the Bullet about its Axis.'

In a subsequent paper Robins explains 'that almost all bullets receive a whirling motion by rubbing against the sides of the pieces they are discharged from'. He also describes some splendid experiments in which 'by means of screens of exceeding thin paper, placed parallel to each other at proper distances, this deflection in question may be many ways investigated. For by firing bullets which shall traverse these screens, the flight of the bullet may be traced out.'

Robins did not provide a drawing of his experiments but Figure 5.1 illustrates the principle. It shows an experiment in which the bullets were fired through two screens and finally hit a wall where the location of the impact was marked.

Fig. 5.1. Sketch of one of Robins's experiments for locating the flight of a musket ball.

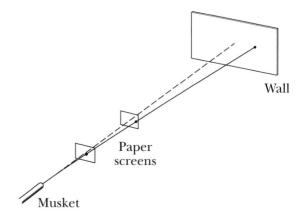

Robins was the first person to understand and measure the effect of spin in producing a sideways force on a rotating object. This effect might properly, therefore, be called the Robins effect. However, it generally goes by the name of the Magnus effect. Heinrich Magnus carried out experiments on rotating cylinders and published his paper 'On the deviation of projectiles, and on a remarkable phenomenon of rotating bodies' in 1852, a century after Robins's work.

Magnus recognizes the contribution of Robins and writes 'Robins, who first attempted, in his *"Principles of Gunnery"* to account for this deviation, thought that the deflecting force was generated by the rotation of the projectile; and at present this opinion is generally accepted.' However, for Robins the existence of the spin was only a plausible assumption. In Magnus's experiments the generation of the spin was controlled and the resulting force was observed directly. His experiments used a rotating cylinder supported on the arm of a torsion balance. Air driven by a fan was directed onto the cylinder and the resulting deflection of the arm observed. In the process he determined the relation of the direction of the force to the spin of the cylinder. Magnus's sketch of his apparatus is reproduced in Figure 5.2. The torsion balance consists of a horizontal beam supported on a wire. The experimental cylinder is supported at the end of the beam (right) with a counterweight at the other end. The cylinder could be made to rotate in its bearings (see insert) and the horizontal airstream from the fan then produces a sideways force on the cylinder and a twisting of the beam about its wire support. This allowed the magnitude and the direction of the force on the spinning cylinder to be measured.

Fig. 5.2. Magnus's diagram of the apparatus he used to study the deflection of a spinning cylinder hung on the arm of a torsion balance and placed in the airstream produced by a fan.

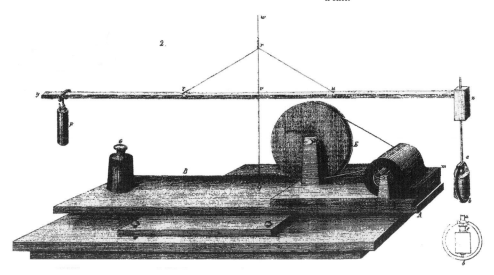

Heinrich Magnus (1802–1870)

Magnus also attempted an account of the fluid dynamics of the effect. Although this is now seen to be inadequate, it is probably the reason that Lord Rayleigh wrote of the effect: 'The true explanation was given many years ago by Prof. Magnus' and, in turn, this comment is possibly the reason for the effect to be named after Magnus. This book will share the honour between the two men and use the term Magnus–Robins effect.

The conventional, incorrect, explanation of lift

The physical processes underlying the lift due to spin were finally clarified after Prandtl provided the understanding of flow over bodies in terms of the boundary layer, as described in the previous chapter. However, the explanation of the Magnus–Robins effect usually given in sports books and by sports commentators is incorrect.

The usual account explains that the air is accelerated on the side of the ball that is moving with the airflow, and is decelerated on the other side where the surface of the ball is moving against the flow. Bernoulli's relation between speed and pressure, described in the previous chapter, is then invoked to show that on the low-speed side the pressure will be higher, resulting in the sideways force.

Now Bernoulli's relation comes from a simple equation of motion for the fluid which assumes that the change of flow speed along a streamline is produced by the pressure gradient along the streamline. However, the spin of the ball only directly affects the flow speed in the boundary layer, and the resulting changes in air speed are driven not by a pressure gradient but by the viscous force through which the ball's surface drags the air. Outside the viscous boundary layer, the spin has no direct effect on the airflow. It is, therefore, not correct to use Bernoulli's relation to deduce a lift force from a pressure imbalance.

Physics of the Magnus–Robins effect

First we recall that the airflow in the boundary layer is slowed as the air flows over the ball. As a result the flow does not reach the back of the ball and separation from the surface occurs. With a spinning ball the viscous force in the boundary layer pushes the air more strongly on the side where the ball's surface is moving with the airflow. This causes the separation to be delayed and allows the flow to continue further round the ball's surface. On the other side the separation is brought forward. This asymmetry distorts the overall flow pattern, as illustrated in Figure 5.3, where it is seen that the flow and the wake are now deflected to one side.

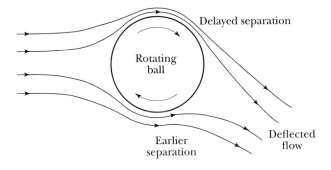

Fig. 5.3. Rotation of the ball leads to an asymmetric separation of the airflow.

That this distortion of the flow pattern produces a force follows from the fact that the deflected flow behind the ball carries sideways momentum. This means that the ball must be subject to a reaction force through which it receives an equal and opposite momentum as shown in Figure 5.4. An alternative, simpler, way of seeing how the force arises is to think of the air as bouncing off the ball, pushing the ball to one side.

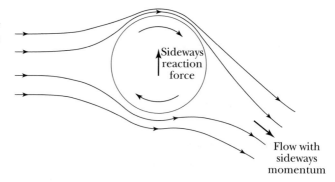

Fig. 5.4. The airflow is deflected by spin, with a sideways reaction force on the ball.

Sideways reaction force

Flow with sideways momentum

The above account describes the behaviour from the ball's 'point of view', and we now need to translate this to find what we shall observe watching the ball in flight. In the above figures the air is taken to be moving to the right, so we would see the ball moving to the left. The force on the ball would therefore deflect it, as shown in Figure 5.5(a). The opposite spin moves the ball in the opposite direction, as shown in (b).

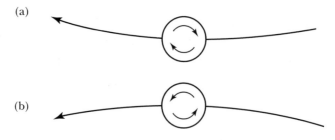

Fig. 5.5. An illustration showing the relation of the direction of the deflection of the ball to the direction of the spin.

(a)

(b)

When the ball is struck with a club it is given back-spin, and when viewed from the side this corresponds to Figure 5.5(a). It is seen that the force arising from the spin will give the ball an upward lift.

A simple experiment

The Magnus–Robins force on a golf ball is recognized from its effect on the ball's flight. The spin is not observed directly and the connection between the spin and the force is made by reference to fairly complicated experiments or to a rather theoretical account of the physics involved. However, it is possible to reproduce the effect in a simple experiment.

Roll a sheet of writing paper into a loose cylinder, holding it in shape with adhesive tape. Wind a length of thread around the centre of the cylinder and, holding the end of the thread, let the cylinder drop. The fall of the cylinder makes it spin and when it falls away from the thread the spinning cylinder moves sideways, demonstrating the Magnus–Robins effect and confirming its direction.

How much lift?

As we have seen, the physics of the lift on a golf ball involves quite complex processes. At speeds of interest the boundary layer becomes turbulent, and the different separations on the two sides lead to a deflected turbulent wake. There is, therefore, no simple calculation of the magnitude of the consequent lift and its variation with the spin and speed of the ball. Our knowledge of these subjects comes, therefore, from experimental measurements.

The classic work is that of Bearman and Harvey[1], who made measurements of the drag and lift in a wind tunnel. In order to achieve the appropriate conditions

[1] P.W. Bearman and J.K. Harvey, 'Golf ball aerodynamics', *Aeronautics Quarterly* 27 (1976), p. 112.

they used models of golf balls which were 2.5 times the normal size and scaled the results to the size of a golf ball. The models were constructed as hollow shells and each shell was split in two to accommodate a motor and bearing assembly on which the ball revolved. The details of the construction are shown in Figure 5.6.

Fig. 5.6. Bearman and Harvey's diagram of the ball model they used to measure the forces on a spinning ball.

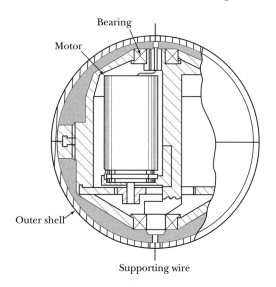

The ball was suspended by a thin wire and a second wire was attached to the underside of the ball, passing through the wind tunnel floor and carrying a weight to keep the spin axis vertical. The two wires also served to supply current to the motor inside the ball. The upper support wire was attached to a strain gauge that measured the lift force of the ball. Since the spin axis was vertical the 'lift' force was actually sideways.

Bearman and Harvey's results have been used to calculate the variation of the lift force for typical golfing conditions. The spin rate is usually given in revolutions per minute, but since the ball is only in flight for seconds, we shall use the more intuitive measure of revolutions per second. A typical spin rate in a drive is 60 revolutions per second and Figure 5.7 gives the variation of the lift force with speed for this spin rate. Since drives can give the ball speeds up to around 160 miles per hour it is seen that it will be quite usual for the lift force to exceed the weight of the ball. This means that in the early part of the ball's flight, before it has slowed significantly, the ball behaves as though it has negative weight.

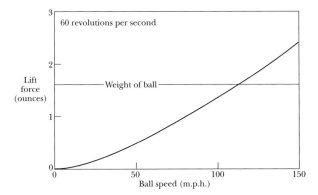

Fig. 5.7. Variation of the lift force with ball speed for a ball spinning at 60 revolutions per second.

The spin given to the ball depends on the loft and speed of the club. For a driver clubhead speed of 100 miles per hour, the ball leaves the club with a spin in the range 30–70 revolutions per second, depending on the loft, and with a speed typically around 140 miles per hour. Figure 5.8 shows how the lift force on the ball increases with spin at this speed. It is clear that the range of a drive will depend on the spin and this will be analysed in Chapter 7.

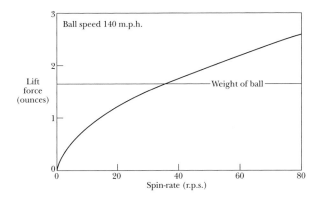

Fig. 5.8. Variation of the lift force with spin rate for a ball travelling at 140 miles per hour.

The energetics of spin

A spinning ball has rotational kinetic energy. A golf ball rotating at the typical rate of 60 revolutions per second has a rotational kinetic energy of 0.6 joules. This is quite small and to put it in perspective we note that the kinetic energy of a ball moving at 100 miles per hour is 46 joules.

The lift that the spin gives to the ball provides a vertical force and this force changes the energy of the ball's vertical motion. When the ball leaves the clubhead in a drive, a typical value for the increase in the energy of the ball's vertical motion due to the vertical lift force is around 4 joules in the first second alone. Since this is much larger than the kinetic energy of the spin, it is clear that it is not the spin energy which provides the energy to lift the ball. So where does the energy come from?

The solution of this problem is quite surprising. Because the 'lift' force is at a right angle to the direction of the ball's velocity, the total energy this force supplies to the ball is zero. The energy transmitted to the ball by the vertical component of the lift force is exactly balanced by the energy extracted from the ball's horizontal motion. As shown in Figure 5.9 the horizontal component of the lift force is in the opposite direction to the horizontal velocity and consequently the ball is slowed, the decrease in energy being taken up by the vertical motion.

Fig. 5.9. The lift force has a horizontal component that takes energy from the ball to provide the energy used to lift the ball vertically.

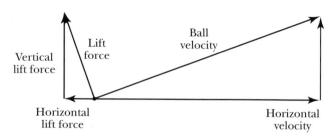

So, what the spin does is to deflect the ball's motion without supplying any energy. We can think of its role as being similar to that of a rudder turning the motion of a boat. In fact, in a drive the ball retains most of its spin to the end of its flight, and its rotation on reaching the ground is typically around 75% of its initial rate.

6 SOME THEORY

The previous two chapters have described the physics underlying the drag and lift on the ball during its flight. In order to fully understand their role in determining the trajectory and range of the ball we need to look in more detail at the basic theory.

It is perhaps helpful to recall here that in the calculations SI units—metres, kilograms, and seconds—are used, but in discussing the results they will be converted to the more familiar units used throughout the book, miles per hour and so on.

The Reynolds number

In 1883 Osborne Reynolds published an account of his brilliant experiments on the flow of water along a pipe. He found that as the speed of the water is increased there is a transition from smooth to turbulent flow. He further explained that the onset of the turbulence with fluids having different densities, viscosities, and speeds flowing in pipes having different diameters is determined by a single pure number, now called the Reynolds number, defined by

$$\text{Reynolds number} = \frac{\text{density} \times \text{speed} \times \text{diameter}}{\text{viscosity}}.$$

It is found that in round pipes the flow is smooth for Reynolds numbers below 2100 and turbulent for values above 4000. Between these values the flow is irregular. The expression for the Reynolds number is simplified further by using the so-called kinematic viscosity which is the viscosity divided by the density. We then have

$$\text{Reynolds number} = \frac{\text{speed} \times \text{diameter}}{\text{kinematic viscosity}}.$$

It turns out that the Reynolds number is also a fundamental factor in the movement of a sphere through a fluid, such as a golf ball through the air. In this case the speed and diameter in the above equation are the sphere's speed and diameter.

Drag coefficient

In simple models of the drag on a sphere it is found that the drag is proportional to the density of the fluid, to the cross-sectional area of the sphere, and to the square of the ball's speed, and the drag force is conventionally written as

$$\text{drag force} = \frac{1}{2}C_D \times \text{density} \times \text{area} \times (\text{speed})^2.$$

The density of the air at 20°C is 1.2 kilograms per cubic metre and the cross-sectional area of the golf ball is 1.43×10^{-3} square metres. C_D is a number called the drag coefficient and this coefficient depends only on the Reynolds number. So once the dependence of C_D on the Reynolds number has been determined for one sphere we can calculate the drag force for all similar spheres of any diameter in any fluid for a range of speeds. We have to say 'similar' spheres because, if the surface of the sphere is not smooth, the form of C_D depends on the nature of the surface, for example, the dimpled surface of a golf ball.

Figure 6.1 shows values of C_D measured for smooth spheres over a wide range of values of the Reynolds number. It also gives values for spheres of prescribed degrees of surface roughness and for a dimpled golf ball. It is seen that for each case there is a rapid fall in C_D at a critical value of the Reynolds number. This corresponds to the drop in the drag force introduced in Chapter 4.

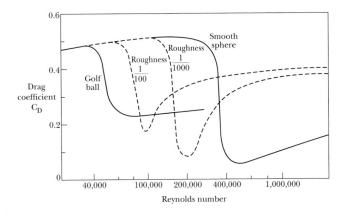

Fig. 6.1. Graphs of the drag coefficient plotted against the Reynolds number for spheres with different degrees of surface roughness and for a dimpled golf ball.

The value of the critical Reynolds number falls as the surface roughness is increased and is about eight times lower for a golf ball than for a smooth sphere. The relationship between the Reynolds number and the golf ball's speed is obtained by substituting the kinematic viscosity of air (1.51×10^{-5} metres^2/second at $20°C$) and the golf ball's diameter (4.27×10^{-2} metres) into the above equation for the Reynolds number to obtain,

$$R = 2830 \times (v \text{ in metres per second})$$

or

$$R = 1260 \times (v \text{ in miles per hour}).$$

For golf ball speeds of interest, say 60–160 miles per hour, the Reynolds number lies in the range 75,000–200,000, and this is seen to fall in the 'plateau' of the C_D curve, with C_D having a value around 0.25.

Lift coefficient

The lift force that occurs with a spinning ball is dealt with in a similar way to the drag force, and a lift coefficient C_L is defined by the equation

$$\text{lift force} = \frac{1}{2}C_L \times \text{density} \times \text{area} \times (\text{speed})^2.$$

Using SI units the equations for the drag and lift forces give the force in newtons, which is the SI unit of force. The force is converted to ounces weight through the relation

$$1 \text{ newton} = 3.6 \text{ ounces}.$$

As described in Chapter 5 the drag and lift coefficients were measured in wind tunnel experiments by Bearman and Harvey. Their measured values of C_L for a Reynolds number of 100,000 are given in Figure 6.2. They are plotted against another pure number, the ratio of the ball's surface rotational speed, v_s, to its directed speed, v. The rotational surface speed is obtained simply by multiplying the ball's spin rate in revolutions per second by its circumference. For example, at a spin rate of 60 revolutions per second the ball's surface speed is 18 miles per hour and for a ball speed of 100 miles per hour this gives $v_s/v = 0.18$.

Fig. 6.2. Bearman and Harvey's measured values of the lift coefficient of a golf ball plotted against the ratio of its surface rotation speed to its directed speed (Reynolds number = 100,000).

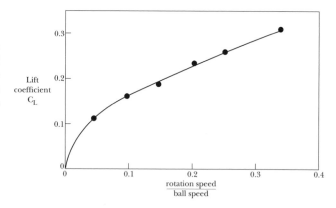

From Figure 6.2 the corresponding value of C_L is 0.21. We can now use the equation above to calculate the lift force for this case. Substituting for the air density and the balls cross-sectional area gives a lift force of 0.36 newtons, which is 1.3 ounces, a little less than the weight of the ball (1.62 ounces).

Drag and spin

The graph of drag coefficient C_D against the Reynolds number in Figure 6.1 was for a ball without spin. We do not expect spin to have a large influence on the drag but Bearman and Harvey did find some variation of C_D with the spin ratio v_s/v, as shown in Figure 6.3.

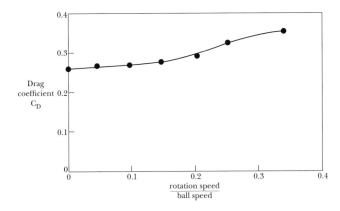

Fig. 6.3. Bearman and Harvey's measured values of the drag coefficient of a golf ball plotted against the ratio of its surface rotation speed to its directed speed (Reynolds number =100,000).

Taking the same example as above, the ball having a spin of 60 revolutions per second and a speed of 100 miles per hour, with $v_s/v = 0.18$, we find $C_D = 0.29$. Using the drag force equation we then obtain a drag force of 0.50 newtons, which is 1.8 ounces, roughly the weight of the ball.

Having introduced the basic procedures for calculating the aerodynamic forces on the ball, we can now proceed to look at their role in determining the trajectory of the ball and consequently its range.

Calculation of the trajectory

At each instant during its flight, the ball experiences a force which depends on its speed and direction. Since these are changing continuously with time, so is the force.

The ball responds to the force, accelerating in the direction of the force according to Newton's law of motion

$$\text{acceleration} = \frac{\text{force}}{\text{mass of ball}}.$$

This equation enables us to calculate the changes in the ball's speed and direction throughout its flight.

In the previous sections we saw how to calculate the aerodynamic forces of drag and lift. The motion of the ball is determined by these forces, together with the acceleration produced by the force of gravity, and the geometry of the forces is shown in Figure 6.4. The drag force is in the direction opposite to the direction of the ball and the lift force is perpendicular to this direction. Because the ball has back-spin the lift force on the ball is upwards, with a small horizontal component.

Fig. 6.4. The forces acting on the ball during its flight.

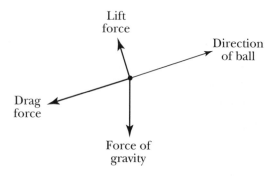

The drag and lift forces change considerably during the flight of the ball but the gravitational force is constant, producing a dowward acceleration of 9.81 metres per second per second, which is 22 miles per hour per second.

The process of calculating the trajectory of the ball requires solving Newton's equation separately for the horizontal and vertical parts of the motion. It is seen from the figure that both the drag force and the lift force act partly in each direction. Since we have formulas for the drag and lift we now have the tools to calculate the trajectory by solving the two equations

$$\frac{\text{rate of change}}{\text{of horizontal speed}} = \frac{\text{horizontal drag} + \text{horizontal lift}}{\text{mass of ball}}$$

and

$$\frac{\text{rate of change}}{\text{of vertical speed}} = \frac{\text{vertical drag} + \text{vertical lift}}{\text{mass of ball}} + \frac{\text{gravitational}}{\text{acceleration.}}$$

The solution of these two equations gives the variation of the ball's speed and direction during its flight, and, having determined this at each time, it is then possible to calculate the ball's trajectory and range.

The equations are quite complicated and we have to resort to the computer to solve them. In the next chapter we shall calculate the range of a ball for different cases and use the results to understand the underlying factors that influence it. The calculations use the form of the drag and lift coefficients given in Figures 6.2 and 6.3 with a small modification that increases the calculated ranges slightly, bringing them into line with present experience.

7 RANGE

The previous chapters have examined the physics of the drive and the aerodynamics of the flight of the ball. This brings us to the purpose of the drive—to hit the ball as far as possible. It is likely that readers found the physics of the swing and impact more complex than they had imagined, but the outcome can be concisely described. The ball leaves the tee with a certain speed, spin, and launch angle, and these quantities determine the trajectory and range of the ball.

The flight of the ball looks quite simple, but behind this appearance is a subtle interplay of three forces—the downward force of gravity, the air-drag, and the lift arising from the spin. It turns out that in a typical drive all three forces are of comparable magnitude.

Using the theory described in the last chapter, the trajectory and range of a drive can be calculated for each case and this will allow us to answer a number of interesting questions. For example, how does the range depend on clubhead speed and what is the optimum club loft for a player with a given clubhead speed?

To obtain a full answer to these questions we have to include an estimate of the distance the ball travels after reaching the ground. The ball initially bounces and then rolls, the two together constituting the run. The total range is then the length of the flight, technically called the carry, plus the length of the run.

When we have examined these basic aspects of the drive we shall ask two further questions. How sensitive is the range to the choice of loft, and is there an advantage from hitting the ball on the rise? Finally we shall see how the flight and range are affected by the wind and by variations in the atmospheric pressure.

The best place to start our study of the flight of the ball is with the idealized case where the effect of the air is neglected.

Idealized flight

It was in the seventeenth century that the Italian astronomer and physicist Galileo determined the shape of the curved path travelled by projectiles. He recognized that the motion could be regarded as having two parts.

From his experiments he had discovered that the vertical motion of a freely falling object has a constant acceleration and that the horizontal motion has a constant velocity. When he put these two parts together and calculated the shape of the projectile's path he found it to be a parabola.

Following Newton, we would now say that the vertical acceleration is due to the earth's gravity, and call the acceleration g. Everyone realized, of course, that Galileo's result only applies when the effect of the air is unimportant. It was obvious, for instance, that a feather does not follow a parabola. The forces from the air increase with the speed of the projectile and are negligible at very low speeds. In the context of golf this means that the flight of slow, short-range shots will be almost parabolic.

For these parabolic trajectories the range depends only on the launch angle and speed. The maximum range for any launch speed is obtained for a launch angle of 45°. This is illustrated in Figure 7.1 which shows the trajectory for 45° together with those for 30° and 60°, all for the same initial speed. The 30° and 60° cases both have a range that is 13% less than that for a 45° launch.

To better understand this we look at the velocity of the ball in terms of its vertical and horizontal parts. The distance the ball travels before returning to the ground is calculated by multiplying its horizontal velocity by the time it spends in flight. If the ball is hit at an angle higher than 45° its time in flight is increased, but this is not sufficient to compensate for the reduction in horizontal velocity, and the range is reduced. Similarly, at angles below 45° the increased horizontal velocity does not compensate for the reduced time in flight. In the extreme cases this becomes quite obvious. For

Fig. 7.1. With no air-drag the ball flies in a parabola, the shape depending on the launch angle. The figure shows the trajectories of three balls launched with the same speed but at different angles.

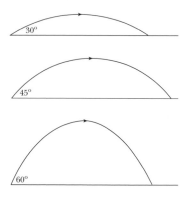

a ball launched vertically the range is zero, and a ball launched horizontally does not leave the ground.

These effects are brought out more fully in Figure 7.2 which shows the horizontal velocity and the time of flight for all launch angles. When multiplied together they give the range shown, with its maximum at 45°.

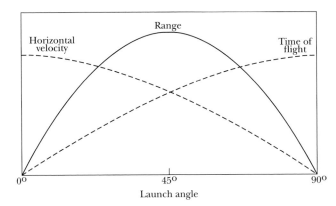

Fig. 7.2. For parabolic trajectories the range is obtained by multiplying the constant horizontal velocity by the time of flight. For balls launched with the same speed both of these depend on the launch angle. As the launch angle is increased the horizontal part of the velocity falls and the time of flight increases, giving a maximum range at 45°.

In the idealized case in which the effect of the air is neglected, the range of the ball is proportional to the square of its launch speed. So a ball with twice the initial speed would travel four times as far. This is illustrated in Figure 7.3 which shows the trajectories of two balls, both launched at an angle of 20°, but one having an initial speed of 80 miles per hour and the other of 160 miles per hour.

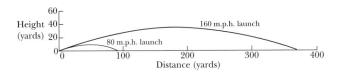

Fig. 7.3. Trajectories of two balls launched at 20° with no air-drag. One has an initial speed of 80 miles per hour, the other 160 miles per hour.

The effect of drag

The trajectories of a golf ball for the two launches described above have been recalculated to include the effect of drag using the procedures outlined in the previous chapter. In Figure 7.4 these trajectories are compared to those for the idealized, no drag, cases. It is seen that drag has only a moderate effect for the slower ball but that the faster ball suffers a very large reduction in range. Whereas without drag the faster ball had a range four times that of the ball with half the launch speed, air-drag has reduced this ratio to about three.

Fig. 7.4. Trajectories of two balls launched at 20° with speeds of 80 miles per hour and 160 miles per hour, now including the effect of air-drag. The trajectories with no drag are shown for comparison.

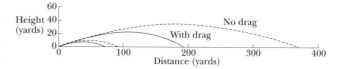

The effect of drag also depends on the launch angle and alters the optimum launch angle for maximum range. By taking a typical drive speed for the ball of 130 miles per hour, the range has been calculated for all launch angles, with and without drag, and the results are compared in Figure 7.5.

Fig. 7.5. Dependence of range on launch angle for a ball having an initial speed of 130 miles per hour. The range with drag is compared to that calculated neglecting drag.

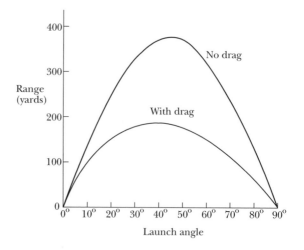

The reduction in range due to drag is substantial, the maximum range being reduced from 376 yards to 189 yards. It is also seen that the launch angle for maximum range has been reduced from 45° to 40°. This is due to the longer time of flight at higher angles giving the drag force longer to slow the ball. For example, with a 60° launch, the ball's horizontal velocity has been reduced to 36% of its initial value by the end of the flight.

Drag plus lift

We now turn to the realistic case where we need to consider both drag and the lift that arises from the spin of the ball. The basic effects of the lift are brought out in Figure 7.6. This shows the modification of the trajectory for the case considered earlier with a 160 mile per hour launch speed and a 20° launch angle, the ball now being given a spin of 60 revolutions per second. There is an obvious change in the shape of the trajectory, the lift force taking the ball upward to more than twice the height of the trajectory with drag alone. The other clear feature is that the lift provides a significant increase in the range.

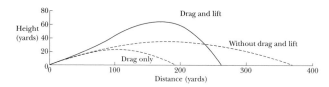

Fig. 7.6. Trajectory of a ball launched at 20° with a speed of 160 miles per hour and a spin of 60 revolutions per second, showing the effect of lift. The trajectories without drag and lift and with drag only are shown for comparison.

Another effect is brought out by again examining the variation of range with launch angle for the case discussed earlier where the ball had an initial velocity of 130 miles per hour, but now with the ball given a spin rate of 60 revolutions per second. The result is shown in Figure 7.7. The introduction of spin has produced several changes. The most important is that the lift produced by spin has increased the maximum range. There is also a substantial lowering of the launch angle at which maximum range is obtained, the angle now being 19°.

Fig. 7.7. Dependence of range on launch angle for a ball launched with a speed of 130 miles per hour. The range with both drag and lift is compared to that with drag only.

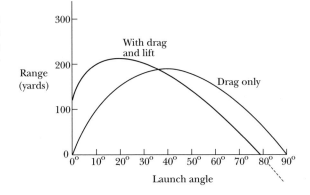

The explanation for the lowering of the optimum launch angle is brought out in Figure 7.8, which compares the trajectories for a 30° and a 60° launch angle. With no air effect these two cases would have the same range. At 30°, the lift force mainly acts vertically, prolonging the flight and giving a longer range. At 60°, the term lift force is hardly appropriate, the force being mainly horizontal. The resulting loss of vertical lift then leads to a much reduced range.

Fig. 7.8. At lower launch angles the lift force acts mainly vertically, actually providing lift, but at higher angles the lift is reduced, the force then acting mainly horizontally.

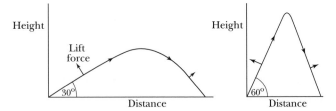

Two other rather surprising results are apparent from Figure 7.7. The first is that at a zero degree launch angle the calculation gives a range of 120 yards. The reason is that the spin initially provides a lift force that is greater than the weight of the ball and as a result the ball rises after its departure. As the ball's velocity is reduced by drag the lift force decreases and the ball falls back to earth.

The other interesting, even though impractical, result of the calculations is that for a 90°, that is vertical, launch the range is negative. This is because the 'lift' is perpendicular to the ball's velocity, and its direction is now essentially horizontal. On the ball's upward journey the lift force is backward and the ball is given a negative horizontal momentum. On its downward path the lift force is forward and this horizontal momentum is removed. However, for almost all of the ball's flight the ball is moving backwards, giving a negative range.

The first general conclusion from these calculations is that air-drag causes a large reduction in the range. The second is that spin and the lift that it provides are very important, and the third is that the combined effects of drag and lift reduce the optimum launch angle, from 45° to 19° in the example we have taken

However, although the example we have used has allowed us to uncover some basic physics, it involved an unrealistic assumption that is somewhat misleading. In the calculations, the ball's velocity and spin were taken to be the same for all launch angles. However, for a given clubhead speed, varying the loft of the club to vary launch angle also changes the ball's speed and spin.

Assuming the clubhead is moving horizontally when it strikes the ball, increasing the launch angle implies increased club loft and higher spin. Clearly a club with zero dynamic loft would launch the ball at zero angle and with zero spin, not the 60 revolutions per second assumed in the above calculations.

Optimum loft

In maximizing the range of a drive there are basically three quantities the player can control—the clubhead speed, the loft of the driver, and the angle of the clubhead's motion at impact. If we assume that each player has a maximum achievable clubhead speed and hits the ball with the clubhead moving horizontally, it then remains to find the loft that will maximize the range he can obtain.

We recall that there are three variables that determine the range—the launch angle, the ball's speed, and its spin. The equations of mechanics allow us to calculate all of these from the clubhead speed and the loft. The formulas that give these quantities were derived by Penner.[1]

Let us take as an example a player who hits the ball with a clubhead speed of 100 miles per hour using a club giving a coefficient of restitution of 0.8. We can then calculate the launch angle, ball speed, and spin for each value of the loft using Penner's equations, and the results are shown in the graphs of Figure 7.9. Using these results we can calculate the trajectory and range for chosen values of the loft.

However, it is important to remember that the effective loft—the dynamic loft—includes the contribution from the forward bending of the shaft. Information on this bending is rather scarce but for the case we are studying the bending angle would typically be around 4°.

Fig. 7.9. Variation of the ball speed, launch angle, and spin rate with dynamic loft calculated for a clubhead speed of 100 miles per hour.

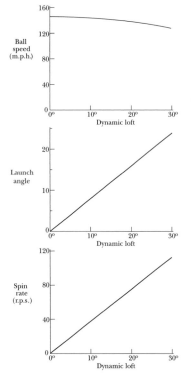

[1] A.R. Penner, 'The physics of golf: The optimum loft of a driver', *American Journal of Physics* 69 (2001), p. 563.

To obtain the club loft, the bending angle has to be subtracted from the dynamic loft.

It is seen that as the loft is increased the ball speed decreases, slowly at small angles but increasingly at higher angles. Both the launch angle and the spin rate increase almost in proportion to the loft.

For each value of loft the corresponding values of launch angle, ball speed, and spin determine the range of a shot. We shall shortly be including the length of the run in the total range but for the present we shall consider only the carry—the horizontal distance the ball has travelled when it first reaches the ground. The calculated dependence of the carry on the loft is plotted in Figure 7.10. The maximum carry for our assumed clubhead speed of 100 miles per hour is 233 yards and this is obtained with a dynamic loft of 15°. For a shaft bending angle of 4° this would mean a club loft of 11°.

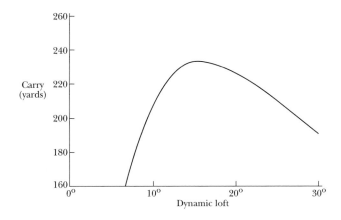

Fig. 7.10. Variation of the carry with dynamic loft, calculated for a ball hit with a clubhead having a speed of 100 miles per hour.

Forces on the ball

In the example we have just considered, the initial drag on the ball is larger than its weight. Because this force is so strong it rapidly slows the ball as shown in Figure 7.11 which gives a graph of the horizontal speed of the ball throughout the flight for a dynamic loft of 15°.

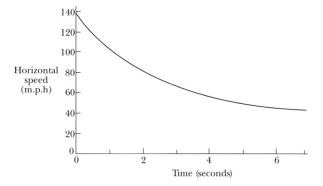

Fig. 7.11. Variation of a ball's horizontal speed during its flight for a ball hit with a clubhead having a speed of 100 miles per hour and a dynamic loft of 15°.

In the case of optimum loft, the initial lift on the ball is 50% larger than the force of gravity and this means that the ball is accelerated upwards. However, the subsequent decrease in the horizontal speed implies a rapid fall in the lift force, and within the first second the vertical lift force has fallen to a value smaller than the force of gravity.

It has been noted earlier that the lift force can have a substantial horizontal component. The geometry of this effect is illustrated in Figure 7.12. The lift force is perpendicular to the direction of the ball and this means that when the ball is rising it produces a horizontal deceleration. In the case with maximum range the horizontal component of the lift force is initially about a fifth of the vertical component. In the graph of horizontal speed in Figure 7.11, it is seen that at the end of the ball's flight the horizontal speed is not falling, the now accelerating horizontal component of the lift force just balancing the horizontal drag. In other cases, with a steeper descent, the ball is actually being accelerated towards the target as it approaches the ground.

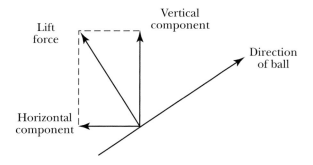

Fig. 7.12. Geometry of the lift force showing how the horizontal and vertical components are related to the direction of the ball.

The run

The carry gives only part of the total range of the ball. When the ball reaches the ground it bounces several times and then rolls along the ground until the friction force between the ground and the ball brings the ball to rest, as illustrated in Figure 7.13. Taken together the travel of the bounces and the roll constitute the run.

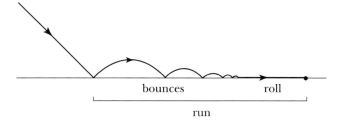

Fig. 7.13. The bounces and the roll constitute the run.

Both the bounce and the roll are very variable quantities. If the ground is wet and soft the ball produces a crater when it first reaches the ground, losing almost all of its energy in the process. The subsequent run is then quite short. In the other extreme, with a frozen or hard dry fairway, the ball will run a long way. In normal circumstances the run constitutes significantly to the overall range and, because balls travelling at a lower angle have a longer run, there is an effect on the optimum loft.

The mechanics of the bounce and the roll in different conditions will be dealt with in Chapter 8. However, there is no point in trying to include the details of the variability of conditions in calculating the effect of the run on the total range, and we shall adopt a simple empirical model that has a physical basis and represents the run quite well for typical conditions.

The underlying physics of the run is essentially the reduction of the kinetic energy of the ball's horizontal motion by the frictional forces until the ball is brought to rest. In fact for the roll part of the run, the expected length is indeed proportional to the ball's kinetic energy at the start of the roll. A reasonable conjecture is that the total run is approximately proportional to the kinetic energy of the ball's horizontal motion when it first reaches the ground.

The kinetic energy of the ball's horizontal motion is $\frac{1}{2}mv_h^2$ where m is the mass of the ball and v_h its horizontal velocity. So the conjecture is that the run is proportional to the square of the horizontal velocity, that is,

$$\text{run} = cv_h^2$$

where c is a constant to be determined.

This formula has been tested by computing v_h for a range of trajectories with a launch angle of $10°$ and comparing the predicted run with the available data. Cochran and Stobbs[2] have given empirical formulas that represent their experimental results and allow a calculation of the run for this case. Direct measurements of

[2] A. Cochran and I. Stobbs, *Search for the Perfect Swing* (Triumph Books, 2005).

the run were made by Williams[3] for a dynamic loft of 12° which also gives a launch angle close to 10°.

Comparison with these results allows a best choice of c and the resulting formula for the run is

$$\text{run} = \frac{v_h^2}{115} \text{ yards,}$$

where v_h is in miles per hour. Figure 7.14 gives a graph of $v_h^2/115$ and compares this prediction of the run with the results of Cochran and Stobbs, and Williams. It is seen that, given the uncertainties introduced by the nature of the fairway, the proposed formula gives a reasonable estimate of the length of the run.

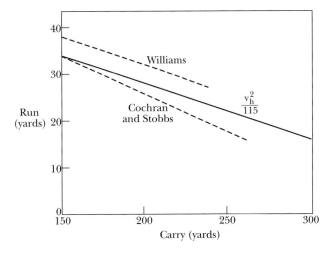

Fig. 7.14. Comparison of the formula for the run with the results of Cochran and Stobbs, and Williams.

It might seem strange that the length of the run is less for longer drives. This is explained by their longer flight times, which give the drag force more time to operate and slow the ball to a lower speed on landing.

[3] D. Williams, 'Drag force on a golf ball in flight and its practical significance', *Quarterly Journal of Mechanics and Applied Mathematics*, XII (1959), p. 387.

Effect of run on total range

Using the model for the run described above the effect of the run on the optimum loft has been calculated for the case analysed earlier. The results are shown in Figure 7.15. It is seen that allowance for the run reduces the optimum loft by about 1°, the lower trajectory providing a longer run, with a run at the optimum of just over 20 yards.

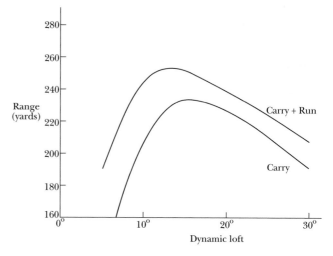

Fig. 7.15. The effect of the run on the optimum dynamic loft for shots with a clubhead speed of 100 miles per hour.

Dependence on clubhead speed

The results of the previous section were for a clubhead speed of 100 miles per hour. It is, of course, of interest to golfers to know the relationship between the optimum loft and the length of the drive.

Players with a clubhead speed of 100 miles per hour have a typical drive range of about 250 yards and their optimum dynamic loft is about 14° implying, say, a 10°

club loft allowing for the bending of the shaft. Let us now look at the results for other clubhead speeds to see the relationship between the optimum loft and the range of the drive.

Figure 7.16 shows how the range for the optimum loft varies with clubhead speed. For a clubhead speed of 70 miles per hour the optimum dynamic loft is 19° and the corresponding range is about 155 yards. With a clubhead speed of 120 miles per hour the range is almost doubled, with an optimum dynamic loft of about 11°. The figure gives an indication of the optimum club loft for players with different ranges and shows that players with shorter drives can gain an advantage by replacing their driver with perhaps a 3-wood.

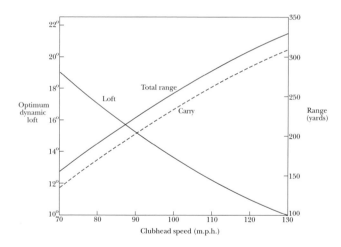

Fig. 7.16. Variation of optimum dynamic loft with clubhead speed. The total range and carry for the optimum loft are also shown.

However, there is a twist to the story. We are probably not very concerned about a few extra yards in range and would not be able to detect such small changes anyway. So how much could the loft differ from the optimum without altering the range by, say, 5 yards? Calculations with a range of lofts produce the results shown in Figure 7.17. It is now seen that using the precise optimum loft is not crucial and, for example, with a 100 miles per hour clubhead speed we find that dynamic lofts between 12° and 16° have drives with ranges within 5 yards difference from that of the optimum 14°.

Fig. 7.17. The insensitivity of the range to variation of loft around the optimum dynamic loft, substantial differences of loft producing a loss of only 5 yards of range.

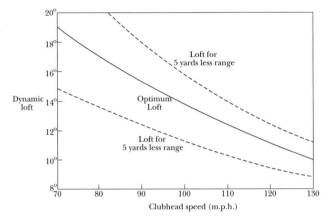

A tricky question

Most of the mechanical aspects of golf are understood in terms of the underlying physics, but one question seems to be unresolved.

It is usual to assume that in a drive the ball is, and should be, struck by the club at the bottom of the swing. In the optimization of the loft angle discussed earlier this condition was applied throughout. With this assumption a given clubhead speed and dynamic loft determine the launch angle. In fact the launch angle is close to 0.8 times the dynamic loft. So the question arises as to what the optimum loft and launch angle would be if they are varied separately. Putting the question another way—at what angle to the ground should the club be moving at impact to obtain the maximum range?

The mechanics of the impact are unchanged if the ball is hit with a rising clubhead, and so for a given clubhead speed and dynamic loft the ball speed and spin are unchanged. It is therefore, straightforward to extend the optimization, now allowing the ball to be struck on the rise, and we shall examine the case of a 100 mile per hour clubhead speed. The geometry is illustrated in Figure 7.18 which defines the rise angle.

The results are shown in Figure 7.19 in the form of a contour map. The axes give the dynamic loft and the rise angle, and the contours are curves of equal total range, carry plus run, plotted at intervals of 10 yards. The results for horizontal impact lie along the line of zero rise angle and the 'highest' point on this line gives a maximum range of 253 yards.

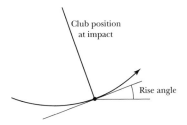

Fig. 7.18. Geometry at impact, showing the definition of the rise angle.

Fig. 7.19. Graph giving contours of equal total range in yards when the dynamic loft angle and rise angle are varied, with a clubhead speed of 100 miles per hour.

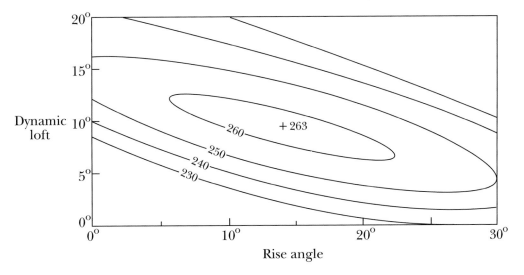

The complete optimization calls for a rise angle of 14°. This gives a maximum range of 263 yards, but under quite different conditions from the optimum for horizontal impact. The optimum dynamic loft is reduced from 14° to under 10° while the launch angle is increased from 11° to a remarkable 22°. The trajectories for the zero rise optimum and the complete optimum are shown in Figure 7.20.

Fig. 7.20. Trajectories for the cases of optimum loft with a zero rise angle and with a complete optimization having a 14° rise angle.

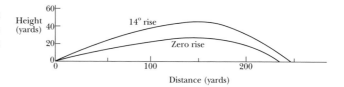

They are quite different, the fully optimized trajectory rising to a much greater height than the zero rise case before reaching the ground 12 yards further on. The gain in distance from allowing a 14° rise angle has come from an increase of more than half a second in flight time, and this has more than compensated for a reduction of lift caused by the spin rate being reduced from 54 revolutions per second to 36 revolutions per second.

So should the ball be struck on the rise, and if so, at what rise angle? The optimum rise angle of 14° for the 100 miles per hour clubhead speed would call for, say, a 3-inch high tee. Whether players could adjust their stance and swing to use this or some lower rise angle is unclear. The matter is further complicated by the dependence of the required rise angle on the clubhead speed. The question of the practical optimum rise angle can only be resolved by experience and experiment, and the author knows of no definitive study of this subject.

The effect of wind

At wind speeds of a few miles per hour the effect of the wind on the ball's trajectory is hardly noticeable, but with wind speeds of tens of miles per hour the changes are obvious. A strong head wind can put a long par-3 hole out of reach whereas a following wind can bring a short par-4 into range.

The force of the wind on the ball is determined by the air speed seen by the ball and this can be calculated by combining the velocity of the wind with the time varying velocity of the ball. With a following wind the ball speed relative to the air is decreased, lowering the drag and increasing the range. With a head wind the relative speed is increased, the resulting higher drag reducing the range.

For low wind speeds the carry is altered by about a yard for each mile per hour of wind speed. At higher wind speeds the effect is larger for a head wind than for a following wind. This is illustrated in Figure 7.21 which gives the trajectories of two balls launched at 130 miles per hour with a dynamic loft of 15° into a 20 miles per hour wind, one with a head wind and the other with a following wind. The carry with no wind would be 209 yards. With a following wind this is increased by 9 yards to 218 yards, whereas with a head wind the drive loses 26 yards reaching only 183 yards. The larger change in the ball's range with a head wind as compared to a following wind is the result of the drag force being not proportional to the speed of the ball relative to the air, but to the square of that speed.

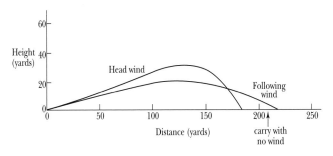

Fig. 7.21. Effect of a 20 miles per hour head wind and following wind on the trajectory and range of a ball launched at 130 miles per hour.

With a head wind the increased relative speed of the ball also increases the lift, producing a higher trajectory as seen in Figure 7.21. Another clear difference between the two cases is the lower angle at which the ball reaches the ground with a following wind. The lower angle is due to the ball's larger horizontal velocity on landing and this has implications for the resulting run.

The trajectory calculations provide the ball's horizontal velocity on landing, v_h. Using this value in the formula for the run then gives an estimate of the run for each wind speed. Taking the same club speed and loft as in the examples given above the length of the run has been calculated and the results are shown in Figure 7.22, which compares the total range including the run with that of the carry alone. It is seen that with a following wind the gain from the increased run is greater than that of the increase in the carry.

Fig. 7.22. The dependence of the carry and total range on the wind speed for balls launched at 130 miles per hour.

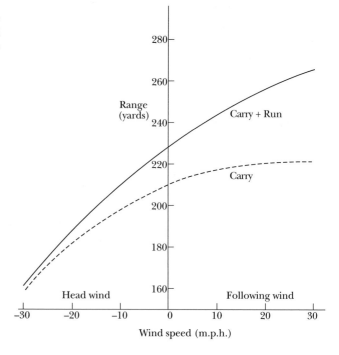

Side wind

While the drag effect of a following and a head wind is intuitively obvious, the mechanics of a side wind is more subtle. Consider first a thought experiment in which an initially stationary ball is subjected to a 20 miles per hour wind. It is straightforward to calculate the motion of the ball and to find how far the ball would move in the flight time of a typical drive, for example, in 6 seconds it would move about 11 yards. However, if we calculate the actual movement of a ball during this typical drive, with a 20 miles per hour side wind, we find that the ball suffers a sideways displacement that is more than twice as large.

The reason for this behaviour is again to be found in the fact that the drag force is proportional to the square of the air speed seen by the ball. This means that the usual drag force on the ball is very much larger than that of the side wind alone. The effect of the side wind is to turn the usual drag force through an angle and the resulting sideways force is then the sideways component of this much larger force.

This can be illustrated by taking the example of a 20 miles per hour side wind acting on a ball moving at 100 miles per hour. We cannot take the ratio of the drag forces at these two speeds to be simply the ratio of the square of the speeds because 20 miles per hour is below the critical speed. So we take the forces from Figure 4.9 and find the ratio to be 12. Figure 7.23 then shows how the deflection of the 12 times larger force leads to a sideways force which is 2.5 times larger than that of the side wind alone.

Fig. 7.23. The way in which a side wind deflects the usual drag force to produce a sideways force much greater than that produced by the side wind alone.

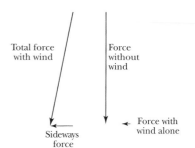

Total force with wind

Force without wind

Sideways force

Force with wind alone

The effect of atmospheric pressure and altitude

The natural variations in atmospheric pressure and temperature were discussed briefly in Chapter 4. The variations in the density of the air are usually in the range ±4%, and produce changes in the drag and lift forces on the ball in the same range.

The effect of differences in atmospheric density on the range of a shot comes from the opposing contributions of drag and lift. An increase in density increases the drag, reducing the range, but also increases the lift, which increases the range. For typical drives at a low launch angle, the two effects roughly cancel and the effect of atmospheric conditions on the range is small. For launch angles over about 10°, change of air density produces a larger effect on the drag than on the lift and so, for example, a reduction in density leads to an increase in range. The change in range is typically 1 yard for each percentage change in air density and so the variations of ±4% in air density imply changes in the range of about 4 yards.

A more significant effect occurs when a game is played at higher altitude with its reduced air density. For example, the reduction in air density at 7000 feet is 21%. The effect of the reduced drag and lift at this altitude has been calculated for different launch angles for a clubhead speed of 100 miles per hour and the result is shown in Figure 7.24. It is seen that at low launch angles the loss of lift leads to a reduction in range whereas at higher launch angles the reduction in drag is more important than the loss of lift and the range is increased by around 20 yards.

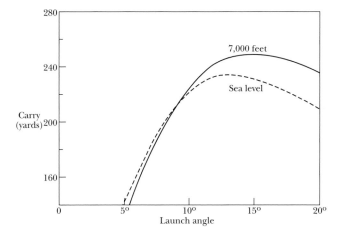

Fig. 7.24. Variation in carry with launch angle for shots driven by a clubhead having speed of 100 miles per hour. The range at an altitude of 7000 feet is compared with that at sea level.

8 BOUNCING & ROLLING

The physics of the interaction of the ball with the ground is quite straightforward, but there is a snag. Both bouncing and rolling are very sensitive to the nature and condition of the ground, and this is so variable that a wide range of behaviour occurs, making its accurate prediction rather pointless. However, for any given set of conditions, Newton's law of motion can be used to calculate the ball's bounce and roll, and the basic features can be illustrated by considering 'typical cases'.

The bounce

When the ball bounces on a grassed surface it loses energy. This occurs partly through the deformation of the grass and the soil, and partly from the friction between the ball and the ground. These effects show themselves in a reduction of the height of successive bounces and in the slowing of the ball's forward motion.

The bounce of the ball depends on two coefficients, the coefficient of restitution, e, which we have already met, and the coefficient of sliding friction, usually represented by the Greek letter μ (myu). We recall that the coefficient of restitution is the ratio of the vertical speed after the bounce to that before the bounce. When a golf ball is launched off the hard surface of a golf club the deformation of the ball is very much greater than that of the clubface, and the resulting coefficient of restitution is close to that of the ball alone. However, when the ball bounces on the fairway or the green, the deformation of the ball is negligible and the bounce is determined by the much less elastic properties of the turf.

It turns out that when a ball is dropped to the ground the ratio of the bounce height to the height from which the ball fell is equal to the square of the coefficient of restitution. So we can determine this coefficient by taking the square root of the ratio of these two heights. For example, if the ball is dropped from a height of 100 inches and bounces to 36 inches, the ratio of the heights is 0.36 and, taking the square root, the coefficient of restitution is 0.6.

We know from experience that when a golf ball is dropped on to a very hard surface it has quite a high bounce. Typically the ball bounces to about two-thirds of the drop height and this gives a value of e of about 0.8. When the ball lands on a fairway which is hard by being frozen or is sun-hardened, a typical value of e is around 0.5, but with a soft surface the value can be as low as 0.1. A further complication is that the coefficient of restitution varies with the ball speed, having a lower value at higher speeds.

When the ball first reaches the ground it slides over the surface, and this results in a horizontal friction force on the ball as illustrated in Figure 8.1. This force is proportional to the vertical reaction force on the ball from its contact with the ground, and the coefficient of sliding friction is defined as the ratio of these forces, that is,

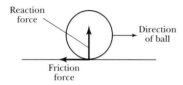

Fig. 8.1. The friction force and reaction force acting on a sliding ball.

$$\text{coefficient of sliding friction} = \frac{\text{friction force}}{\text{reaction force}}.$$

A typical value of the coefficient of sliding friction is 0.4, and this value will be used in the examples we shall consider.

Cleanly hit balls are always given a back-spin, generated by the loft of the club, and they still have back-spin when they reach the ground. As illustrated in Figure 8.2 the friction force then acts both to slow the ball and to reduce the spin.

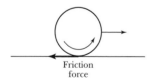

Fig. 8.2. When the ball reaches the ground it has back-spin and the sliding friction force acts to slow both the ball and its rotation.

The bounce from a drive

In the case of a drive, the friction force during the bounce is large enough to reverse the ball's spin and transform the sliding motion into a roll before it leaves the ground. In the process the ball's forward motion is slowed, its forward speed typically being reduced by about a third.

As an example, consider a drive in which the ball has a back-spin of 60 revolutions per second and reaches the ground at 40° to the horizontal with a speed of 55 miles per hour. Taking the coefficient of restitution to be 0.15 the first bounce is calculated to reach a height of 11 inches, the ball then travelling 17 feet before returning to the ground. Before the ball leaves the ground from the first bounce, the sliding has been brought to a halt and the ball is rolling. This means that in the subsequent bounces, the ball has the rolling spin. There is, therefore, no frictional force due to sliding in these bounces and the horizontal speed of the ball is almost unchanged during the bounce. However, at each bounce the vertical speed is reduced by a factor e, the coefficient of restitution. As a result, the time between bounces and the distance travelled in successive bounces are also both reduced by e, and the height of successive bounces is reduced by the square of e. In calculating the second and subsequent bounces the coefficient of restitution has been increased to 0.3, making allowance for the higher value applicable at lower speeds. This means that the height of successive bounces is reduced by the square of 0.3, which is 0.09, so that the height of each bounce would be about an eleventh of that of the previous bounce. This rapid reduction in the height of the bounces means that after a short time bouncing effectively stops and rolling takes over.

These features are brought out in Table 8.1, which gives the length and height of the ball's flight in the sequence of bounces calculated for the case considered above.

Bounce	Length (feet)	Height (inches)
1	17	11
2	5	1
3	1.5	Negligible

Quite often the ball's first bounce is higher than predicted by the basic theory, and this calls for an explanation. The theory of the bounce assumes that the ground is not significantly deformed during the impact of the ball and predicts a low bounce as shown in Figure 8.3(a). However, the pitch-marks left by the ball show that this is not always the case, and Figure 8.3(b) illustrates what can actually happen. On hitting the ground the ball deforms the surface, producing a miniature hill at the front of the ball. It is seen that the ball is then bouncing off an upward slope and this gives the higher bounce.

Fig. 8.3. When the ball lands on a firm surface it takes a low bounce but when the ball makes a pitch-mark it bounces off a deformed surface and has a higher bounce.

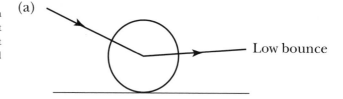

(a) Low bounce

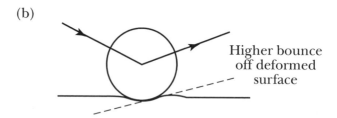

(b) Higher bounce off deformed surface

If the bounce calculation is repeated taking the effective angle of the slope of the ground to be 10° the first bounce is found to be more than three times as high as from flat ground. Consequently, the length of the bounce is increased by 7 feet to 24 feet. Figure 8.4 shows the sequence of bounces for this case. The height of the fourth bounce is only a fraction of an inch, comparable to the height of the grass, and the ball can be taken to be rolling from this bounce.

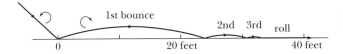

Fig. 8.4. Sequence of bounces with a high first bounce from a pitch-mark.

Bounce with high back-spin

When the ball is hit with a high-number club or a wedge, it is given a high back-spin. This can produce a bounce that is very different from that of a drive. In such cases the friction between the ground and the ball can be insufficient to slow the rapid spinning of the ball enough to turn the sliding motion into a roll.

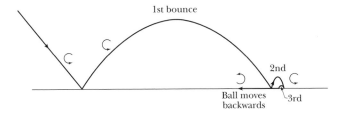

Fig. 8.5. An example of a ball reaching the green with a high back-spin. The first and second bounces carry the ball forward but the back-spin persists, finally taking the ball backward.

An example is shown in Figure 8.5. The ball has reached the ground at 50° to the horizontal with a speed of 50 miles per hour and a back-spin of 180 revolutions per second. It then undergoes three bounces, in all of which the ball is sliding throughout the bounce, and the back-spin, although slowed, is retained. The coefficient of restitution has been taken to be 0.4 and the first bounce takes the ball about 6 feet. The small second bounce is still in the forward direction but the back-spin is sufficient to turn the ball backwards for the even smaller third bounce. After the third bounce the ball still has back-spin and it continues its backward motion skidding on the surface. Friction then removes the skidding and the ball finally rolls to a halt.

Rolling

As we would expect, the frictional drag force on a rolling ball is much less than that on a sliding ball. As with a sliding ball the friction force is measured by a coefficient, called the coefficient of rolling friction, defined by the ratio of the friction force to the weight of the ball, that is,

$$\text{coefficient of rolling friction} = \frac{\text{friction force}}{\text{weight of ball}}.$$

We shall represent this coefficient by the symbol μ_r (myu r). Perhaps surprisingly, the coefficient of rolling friction does not depend significantly on the speed of the ball.

If a golf ball rolls on an extremely hard surface the friction coefficient is very low, around 0.003. So the friction force is then about one three hundredth of the weight of the ball. In this extreme case the friction force is probably influenced by the ball's dimples.

The basic cause of rolling friction on the golf course is the bending of the blades of grass, and generally the longer the grass the greater the friction. On the fairway, a typical value of the rolling friction coefficient is 0.15.

Values on putting greens are generally between 0.05 for a fast green and 0.08 for a slow green, and for a typical value, say $\mu_r = 0.06$, the friction force is about a seventeenth of the ball's weight.

The kinetic energy of the ball is proportional to the square of the ball's speed, v, and the rolling distance d is given by

$$d = \frac{v^2}{90\mu_r} \text{ yards}$$

where v is in miles per hour. Figure 8.6 gives a graph of the rolling distance against ball speed for a typical fairway, taking $\mu_r = 0.15$.

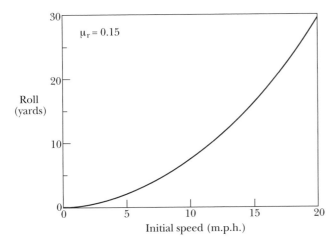

Fig. 8.6. Graph of the rolling distance against ball speed for a typical fairway.

If we take a drive in which bouncing reduces the speed of the ball to 15 miles per hour we see from the graph that the ball would then roll for about 17 yards.

On the green the ball speeds are, of course, generally much lower. For example, with a coefficient of rolling friction of 0.06 on a level green, a 20 foot putt requires an initial speed of only 6 miles per hour.

The 'long putt'

An interesting case of rolling arises when a shot with a driver or fairway wood is mis-hit and the ball is sent along the ground. Although a mis-hit, the ball speed can be quite high and we shall look at a case where the ball starts with a speed of 60 miles per hour.

If the formula given above for the rolling distance, d, is used with $\mu_r = 0.15$ it predicts a roll of 267 yards. This is clearly unreasonable, so what is wrong? First it is necessary to allow for an initial phase where the ball is sliding and the friction coefficient is consequently substantially higher. As the ball slides, its rotation speed increases until the ball is rolling and the friction then drops to the lower value. Assuming no initial spin, the inclusion of this effect reduces the distance to 190 yards, which is obviously still too long.

The resolution of the problem is that at such speeds the air-drag is important even for a roll. Clearly the pattern of the airflow over the ball will be complicated by the ball's proximity to the ground but we can obtain a reasonable estimate of the drag using the normal formula for the drag on a ball in flight. The distance the ball travels then turns out to be a more reasonable, but still quite long, 100 yards. Players are indeed often relieved by the distance they get from such a mis-hit.

The energetics of rolling

When a ball rolls, its kinetic energy has two parts: that due to its forward directed motion and that of its rolling rotation. The slowing of the ball's directed motion is due to the friction force between the ball and the ground as shown in Figure 8.7. As the ball slows, its rotation speed must also decrease, but it is seen from the figure that the horizontal friction force is in a direction to increase, not decrease, the rotation.

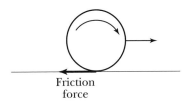

Fig. 8.7. The horizontal friction force will not slow the ball's rolling rotation.

Friction force

This problem is resolved by recalling that there is also a much larger force operating on the ball, namely the reaction to the ball's weight. With a stationary ball this force acts through the centre of the ball but with a rolling ball the force is displaced. We can see how this occurs by looking more carefully at the ball's interaction with the ground. As the ball moves forward it bends the grass as shown in Figure 8.8.

The force on the ball is distributed over its lower surface but the principal effect of rolling is that the resultant reaction force is moved forward as shown in the figure. It is now seen that the overall force is in a direction that slows both the forward motion and the rotation, and takes energy from both.

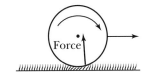

Fig. 8.8. The figure shows how the force on the ball acts to slow both its directed motion and its rotation.

Force

Because rolling implies that the ball's rotation rate is proportional to its speed, the energies of the rolling and forward motion are in a constant ratio. Calculation shows that the forward motion has 5/7 of the total energy and the rotation has the remaining 2/7. The force slowing the ball acts in such a way that it holds these energies in this fixed proportion.

The 'speed' of greens

It seems natural to talk about 'speed' of greens but greens, of course, do not have a speed. What is actually meant is that for a given hit the ball travels faster, and therefore further, on a 'fast' as compared to a 'slow' green. This is clearly related to the coefficient of rolling friction and we would expect that there is a relationship between the distance the ball travels and the

Fig. 8.9. The Stimpmeter.

friction coefficient. This relationship is brought out clearly in the device called the Stimpmeter that is used to measure the 'speed' of greens.

The Stimpmeter was introduced by Edward Stimpson in the 1970s. Stimpson was an able golfer and had been Massachusetts Amateur Champion in 1935. The Stimpmeter, shown in Figure 8.9, is now the standard instrument used throughout the world.

The Stimpmeter consists of an aluminium bar with a v-shaped groove along its entire length. The ball is initially held in a notch 30 inches from the bevelled lower end of the bar, which rests on the ground. The upper end of the bar is slowly raised until the ball falls from the notch. Once released, the ball rolls down the groove reaching the ground with a well-defined speed. The distance the ball rolls on the green then gives a precise measure of the slowing of the ball. It is usual to quote the roll distance, measured in feet, as the speed of the green. Typically a fast green could have a speed of 11 feet and a slow green a speed of 7 feet.

This procedure for measuring the speed of the green is straightforward and gives a result that is easily understood. So, if all we need to know is the speed of the green, then that is the end of the matter. However, if we want to understand the physical processes involved, it turns out that this is more complicated than we might imagine.

The distance the ball rolls is determined by the ball's speed on leaving the Stimpmeter and by the friction force between the ground and the ball. The friction force is measured by the coefficient of rolling friction introduced earlier, and for a given initial ball speed the roll distance is inversely proportional to the coefficient of friction. The complexity arises in calculating the speed with which the ball is launched by the Stimpmeter.

While the energy gained by the ball in falling from the height of the notch of the Stimpmeter is easily calculated, it has to be borne in mind that part of this energy is transferred to the rolling motion of the ball. The rest of the energy is in the directed velocity of the ball. Then, because this motion is along the Stimpmeter and at an angle to the ground, only a part of the ball's velocity is available to carry the ball forward along the ground. Finally, because the speed of the ball changes on reaching the ground there is an associated change in the ball's rotation rate, and this also needs to be taken into account. The sequence of events is illustrated in Figure 8.10.

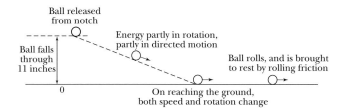

Fig. 8.10. The sequence of events for a ball released from a Stimpmeter.

When all of these stages are included, the calculated horizontal launch speed is 4.2 miles per hour. The resulting relationship between the speed of the green, measured in feet, and the coefficient of rolling friction is

$$\text{green-speed} = \frac{0.60}{\mu_{\text{r}}}.$$

For example, a coefficient of rolling friction of 0.06 would correspond to a green-speed of 10 feet.

9 PUTTING

The putting green provides many challenges for the golfer, and it also raises a variety of questions about the underlying physics.

W e shall look at the speed required to make a successful putt and at the effect of a slope, both
along the line of the putt and across the line. We shall
also enquire about the effect of a wind. But, first things
first, we start by analysing the hitting of the ball with
the putter.

Hitting the ball

Assuming that the ball is hit cleanly with the face of the
putter, there are two factors that decide the accuracy of
the direction of the putt. First, there will be an error
in the direction taken by the ball if the putter is square
to the desired line but the direction of stroke is at an
angle to that line. Second, an error arises if the direction of the stroke is correct but the putter is not square
to the desired line.

Perhaps surprisingly, the mechanics of a mistake in
the line of the stroke is very forgiving. The error in the
angle of the shot is only about a sixth of the error in
the line of the stroke. This is illustrated in Figure 9.1,
which shows the 1° error that results from a stroke that
is 6° off-line.

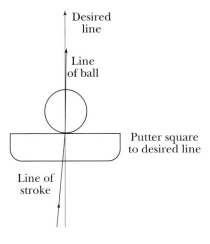

Desired
line

Line
of ball

Putter square
to desired line

Line of
stroke

Fig. 9.1. An error in the line of the
putter stroke leads to a comparatively
small error in the line of the ball.

There are two reasons for this piece of good fortune. The first is that the ball's forward speed is greater than that of the putter by a factor $1 + e$, where e is the coefficient of restitution for the bounce of the ball off the putter face. A typical value of $1 + e$ is 1.7.

The second and more important factor is that when the ball is struck with a motion partly across the face of the putter, the sideways speed of the putter is only partially transmitted to the ball. The reason for this is that during the impact the ball rolls on the clubface as illustrated in Figure 9.2. Calculation shows that the sideways speed of the ball is only 2/7 of the sideways speed of the putter. The reduction of the error is given by ratio of these two factors, 2/7 divided by 1.7, which gives the reduction to 1/6 as quoted earlier.

Fig. 9.2. In a stroke with a misdirected putter, the ball rolls on the putter face and as a consequence the error is reduced.

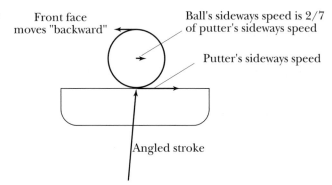

Front face moves "backward"

Ball's sideways speed is 2/7 of putter's sideways speed

Putter's sideways speed

Angled stroke

An error in the alignment of the face of the putter is more serious. The angle at which the ball departs is just a little less than the angle of the putter face. This is illustrated in the diagram of Figure 9.3 where 5° error in the angle of the putter face leads to a putt that is out of line by 4°. This effect places a stringent requirement on the alignment of the putter. An error angle of 1° for the face of the putter would mean that the ball would miss the hole in a 13-foot putt.

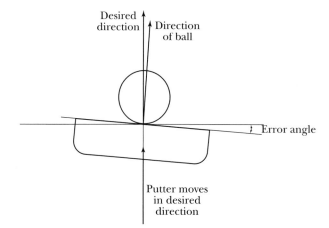

Fig. 9.3. Misalignment of the face of the putter leads to a serious error in the line of the ball.

The simple roll

If the line of the putt lies on level ground, the ball's motion will be determined by the rolling friction. The ball will be slowed at an almost constant rate, the rate depending only on the coefficient of rolling friction. This allows us to calculate the initial speed the ball must be given to produce a putt of a chosen length for a given green-speed. Figure 9.4 gives the initial speed for each putt length taking the coefficient of rolling friction to have the value 0.06, which corresponds to a Stimpmeter green-speed of 10 feet.

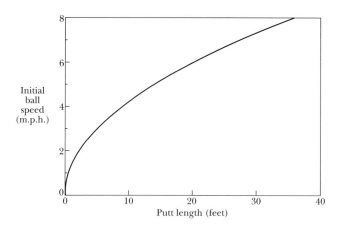

Fig. 9.4. Ball speed required for the ball to reach a given distance on a putting green with a coefficient of rolling friction of 0.06.

Although the slowing takes place at a constant rate, more of the slowing takes place in the latter part of the roll. Figure 9.5 shows how the speed of the ball falls along its path. It is seen that the ball loses the final half of its speed in the last quarter of its roll.

Fig. 9.6. The behaviour of the ball on reaching the hole depends on its speed.

For speeds less than 1 m.p.h. the ball rolls into the hole.

For speeds more than 1 m.p.h. and less than 3 m.p.h. the ball flies and hits the back face of the hole.

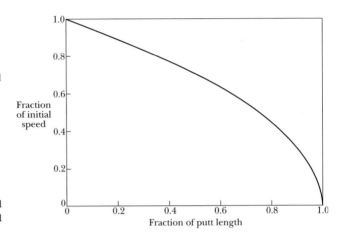

Fig. 9.5. Slowing of the ball along its path.

For balls with speeds greater than 3 m.p.h. there is a critical speed below which the ball bounces off the back edge into the hole.

The critical speed is around 4 m.p.h. depending on the condition of the hole.

Above the critical speed the ball bounces out from the back edge.

At the hole

Assuming the ball is aimed sufficiently well to hit the hole, will it fall in? Let us first consider the case where the ball reaches the 4¼-inch-diameter hole in line with its centre. The trajectory of the ball at the hole can be calculated using the equation of motion together with the theory of the bounces. It turns out that there are several possibilities, illustrated in Figure 9.6.

At speeds less than 1 mile per hour the ball rolls off the edge of the hole. At higher speeds it leaves the ground horizontally and then falls downward. For speeds between 1 and 3 miles per hour the ball is moving slowly enough for it to have fallen below the edge of the hole when it reaches the far side. For speeds greater than 3 miles per hour the ball hits the back edge of the hole and its subsequent fortune depends on the condition of the turf. If the turf is springy the ball tends to bounce back into the hole and speeds around 4 miles per hour are possible. For higher speeds the ball bounces forward, away from the hole, and disappoints the golfer.

If the ball arrives at the hole off-centre, the limitation on its speed is more stringent. A putt that sends the ball close to the edge of the hole must arrive with a very low speed if it is destined to succeed. Figure 9.7 gives a graph of the maximum speed a ball can have if it is to fall into the hole.

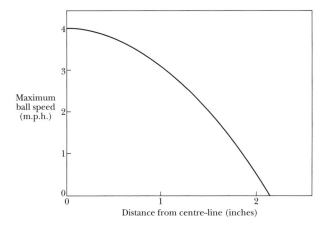

Maximum ball speed (m.p.h.)

Distance from centre-line (inches)

Fig. 9.7. Dependence of the maximum speed for which the ball will enter the hole on the distance from the centre-line of the hole.

Another way of thinking about the effect of ball speed on the likelihood of success is to regard the speed as determining the *effective* size of the hole. For example, a ball travelling at 2 miles per hour will fall into the hole if it is within 1.5 inches of the centre of the hole, so we can say that at this speed the effective hole size is 70% of its actual size. A graph of the effective size against ball speed is obtained by exchanging the axes in Figure 9.7, and the result is shown in Figure 9.8.

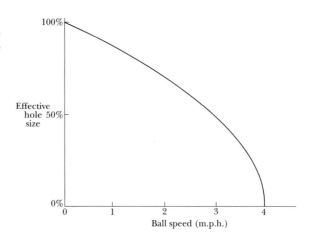

Fig. 9.8. **Fig. 9.8.** Dependence of the effective size of the hole on the speed of the ball on reaching the hole.

Putting on a slope

We will first look at the 'simple' case where the line of the putt is directly up or down the slope. The effect of the slope is to produce a component of the force of gravity along the path of the ball. The force of gravity on the ball is just its weight and for the small slopes that occur on putting greens the gravitational force along the path is proportional to the slope. For a slope of 1 in 10 the force is one tenth of the ball's weight.

Because the friction force is also proportional to the ball's weight it is straightforward to compare the force arising from the slope with the friction force. The ratio is simply

$$\frac{\text{slope force}}{\text{friction force}} = \frac{\text{slope}}{\mu_r},$$

where μ_r is the coefficient of rolling friction. This means that if the slope is greater than μ_r for a downhill putt the slope force will be greater than the friction force and this will produce a resultant force down the slope. The ball will then accelerate rather than slow down. For example, with a typical value of μ_r of 0.06 the ball will accelerate down the slope if the slope is greater than 0.06, which is approximately 1 in 17.

A putt up the slope has less of a problem but there is the danger that, if the slope is greater than the coefficient of rolling friction, a putt that misses the hole will roll back towards the player.

Putting across a slope

There is, of course, a great variety of situations on a putting green with a slope, depending on the position of the ball relative to the hole. To illustrate the complexity of the problem we will take the case where the direction of the hole is directly across the slope. The ball then starts at the same height as the hole.

With a short putt the slope is comparatively unimportant but with longer putts a decision has to be made about both the direction and the speed of the putt. As an example we shall take a 10-foot putt across a slope of 1 in 20 as illustrated in Figure 9.9, the coefficient of rolling friction being taken to be 0.06.

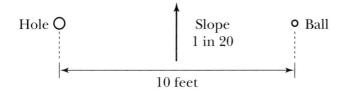

Fig. 9.9. In the example taken, the distance to the hole is 10 feet and the ball has to be putt across a slope of 1 in 20.

A very fast putt aimed directly at the hole would reach the hole but would be too fast to fall into it. The same is true for balls aimed at too small an angle up the slope. On the other hand, if the ball is aimed at too high an angle the ball will come to a halt without reaching the hole, no matter what its speed. These two cases are illustrated in Figure 9.10.

Fig. 9.10. For a putt across the slope to go directly to the hole it requires a speed that is too high for it to enter the hole. If the putt is aimed at too high an angle the ball will stop before reaching the hole.

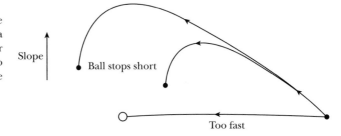

There is a range of trajectories that will succeed in putting the ball into the hole. For each starting angle the ball must be given the correct speed. The smallest angle allowed is one for which the speed of the ball on reaching the hole is just low enough for the ball to enter the hole. For this putt, the ball will need to hit the centre of the hole and the effective hole size is almost zero. As the angle of the putt is increased from this critical value the effective size of the hole increases. The largest successful angle is for a putt which only just reaches the hole. Again this is a difficult putt because for a slightly lower speed the ball will not reach the hole. These limiting trajectories are shown in Figure 9.11.

Fig. 9.11. The limiting trajectories for which the ball just enters the hole.

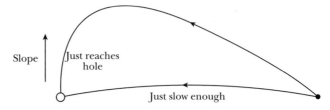

The best putt will lie between these extremes and depends on the type of game being played. In match-play it is often essential to have a successful putt in order to win or halve a hole. It is then, perhaps, worth the risk of a low-angle, almost direct, shot at the hole. Figure 9.12 shows such a trajectory that has a final speed which is half the maximum allowed for the ball to drop, and gives a 75% effective hole size. If the shot fails, the ball will end up 10 feet down the slope – but that is of no consequence.

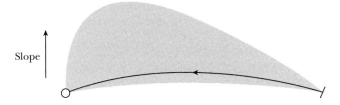

Fig. 9.12. The trajectory of a putt for which the ball reaches the hole with a speed that is half the maximum speed for the ball to drop into the hole and for which the effective hole size is 75%.

Slope

In stroke-play a higher-angle trajectory, with its much lower speed at the hole, is called for to allow for the possible need for a second putt.

Optimal putting

In the previous section we saw that, in putting, two basic situations arise, each calling for a different strategy. The first, which is often faced in match-play, is where the player needs to hole the putt in 1 shot, anything more being worthless. The second situation, which occurs in stroke-play, is where it is necessary to minimize the expected number of putts required to hole the ball.

The basic requirements for the 1-shot case are quite straightforward. The speed given to the ball must be high enough for it to reach the hole but not so high that the ball is too fast to fall into it. Taking a typical rolling friction coefficient of 0.06, the upper and lower limits to the initial speed of the putt have been calculated for different lengths of putt assuming straight putts for which the ball hits the centre of the hole. The results are shown in Figure 9.13. For off-centre putts the band of successful speeds is narrowed.

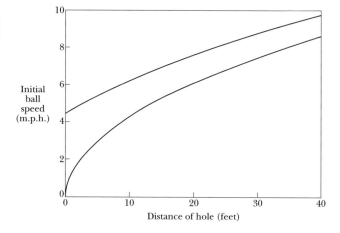

Fig. 9.13. Variation of the upper and lower limits for the initial ball speed with distance from the hole.

The second case, allowing more than one putt, is more complicated. The underlying problem is that it is clearly important that the ball is given sufficient speed to reach the hole, but not be sent so fast that a miss leaves a difficult second putt and the possibility of a third.

The analysis for this case depends on the spread of outcomes for putts aimed to stop at a chosen target distance. This is illustrated in Figure 9.14, which is a graph showing the probability that the ball will end a distance, *d*, long or short, of the target distance.

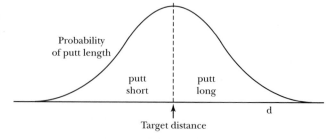

Fig. 9.14. The figure shows the spread of probability of putt lengths around the target distance.

The curve is bell shaped and it is convenient to assume that its shape is that which mathematicians call the 'normal distribution'. The curve has its maximum at the target distance and falls away on either side.

Probability is measured in the range 0 to 1, zero referring to no possibility and 1 to certainty. Figure 9.15 gives an example where the standard deviation is 1 foot. The probability that the actual distance of the putt lies in any range is equal to the area under the curve at that range. For example, the probability that it lies in the range w marked in the figure is given by the area of the shaded region shown.

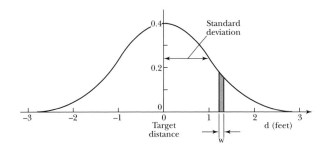

Fig. 9.15. The probability that the ball will stop in the range w is given by the area of the shading.

The degree of expected error is characterized by the width of the curve and this is defined by the 'standard deviation', which is shown in the figure. The definition of the standard deviation is such that 68% of the shots would fall within the standard deviation and 95% would fall within the width of two standard deviations. For a putt of a given length the standard deviation will be different for each player, according to their ability. A typical standard deviation would be about 10% of the target distance, giving a value 3 inches for a 2½-foot putt and 1 foot for a 10-foot putt.

The optimum aiming distance depends not only on the length of the putt but also on the skill of the player. Here we shall illustrate the principles by examining the case of a golfer of average ability.

We start by taking the example of a 10-foot putt. Figure 9.16 gives a typical graph of the probability of success of the first putt, plotted against the target distance beyond the hole. This shows the benefit to be obtained by aiming to hit the ball beyond the hole.

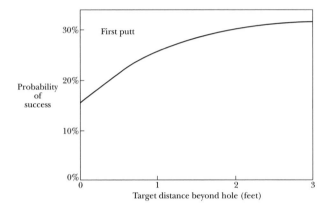

Fig. 9.16. Showing how the probability of success for a 10-foot putt depends on the target distance.

Now we look at the probability that if a second putt is necessary it will be successful, and the result is shown in Figure 9.17. This shows the decrease in success rate with increased target distance for the first putt, implying a corresponding increase in the need for a third putt.

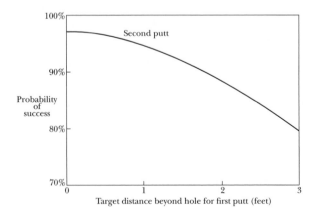

Fig. 9.17. Dependence of probability of success for second putt on the target distance for the first putt for a hole distance of 10 feet.

It is seen that as the target distance is increased there is a trade-off between increasing the probability that the first putt is successful and the increasing likelihood of a need for a third putt. To summarize the results, Figure 9.18 shows how the percentage of cases requiring one, two, and three putts varies with the target distance.

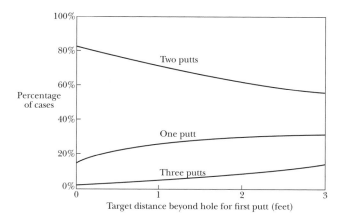

Fig. 9.18. Dependence of the percentage of cases requiring one, two, and three putts on the target distance for a hole distance of 10 feet.

When all of the effects are taken into account it is possible to calculate the average number of putts required for each target distance and the result is shown in Figure 9.19.

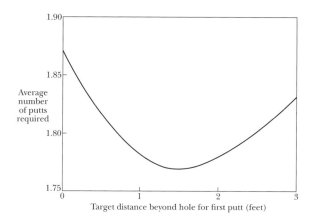

Fig. 9.19. Average number of shots required at each target distance for a hole distance of 10 feet.

The optimum target distance is seen to be about 18 inches beyond the hole and this gives an average of 1.77 putts for the hole.

Similar calculations for 5-foot and 30-foot putts indicate that a target distance of 18 inches past the hole is satisfactory for these distances also, showing that 18 inches is appropriate for a wide range of putt lengths. However, in the case of long putts the average number of putts required is less sensitive to the target distance, largely because success with the first putt is unlikely and consequently it is not so costly if the first putt is short.

Effect of a wind

How much does a wind affect the accuracy of a putt? The first thing to understand is how the wind speed varies with height and in particular the relation between the wind speed at the height of the ball on the green and the wind speed quoted by meteorologists.

In weather forecasts the meteorologists give the wind speed at a height of 10 metres, that is 33 feet, and generally people are not concerned with the speed close to the ground. Aircraft pilots and the crews of sailing boats, for example, are not interested in the wind speed at very small heights. As a result, typical graphs of the dependence of wind speed with height generally assume that the speed is negligible close to the ground. Figure 9.20 gives a typical graph for the case where the wind speed at a high level is 10 miles per hour.

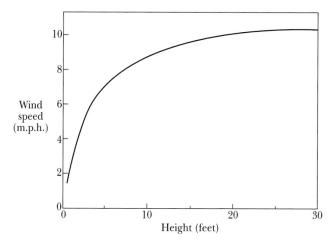

Fig. 9.20. Typical graph of variation of wind speed with height for a high-level wind speed of 10 miles per hour, the graph incorrectly indicates a speed approaching zero at ground level.

However, in putting the golfer is interested in the wind speed an inch above the ground, and we know that the wind speed is not zero there because we see leaves being whisked along by the wind. Measurements of the wind speed at a height of 1 inch show that, far from being zero, the speed is a substantial fraction of the high-level wind speed. The fraction is quite variable but a typical value is around 40% of that at 10 metres. So for the example given above, with a high-level wind speed of 10 miles per hour, the speed at the level of the ball would be about 4 miles per hour, as shown in Figure 9.21.

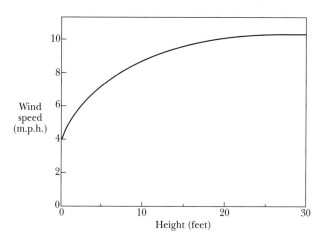

Fig. 9.21. Measurements show that there is a substantial wind speed at ball level. The graph gives a typical variation of wind speed with height for a high-level wind speed of 10 miles per hour.

To calculate the effect of the resulting air-drag on the ball during a putt we need to know how the drag on a ball in contact with the ground depends on its speed relative to the air. We do not have a measurement of the corresponding drag coefficient, C_D, but we can obtain an estimate by taking it to be the same as that for a ball above the ground. At the speeds of interest this gives a value of $C_D = 0.4$.

In calculating the trajectory of the ball we have to include the effect of both the rolling friction and the air-drag due to the wind. The deceleration due to friction will be taken to be that of a green with a Stimpmeter speed of 10 feet. As an example we shall examine the case of a 10-foot putt and see how the path of the ball is affected by a side wind perpendicular to the intended path of the ball.

Fig. 9.22. Trajectories of ball aimed at hole at a distance of 10 feet for several ground level wind speeds.

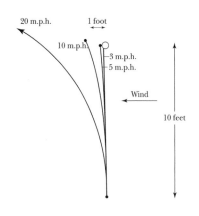

Figure 9.22 shows the paths for wind speeds 3, 5, 10, and 20 miles per hour at ground level. The two lowest speeds correspond to a gentle breeze at a higher level, the 10 miles per hour case would be a strong breeze, and 20 miles per hour would require a strong gale.

Rather surprisingly a ground wind speed of only 4 miles per hour would deflect the ball enough to miss the hole. At 10 miles per hour the ball misses by more than a foot. The 20 mile per hour case is particularly interesting, the force of the wind just exceeds the friction force and the ball continues to roll away. However, this corresponds to a strong gale and the player would be better off in the clubhouse.

Figure 9.23 indicates more directly how the wind affects the putting. It shows how the deflection of the ball over a 10-foot putt depends on the ground wind speed. For wind speeds less than 2 miles per hour the effect is unimportant. At this speed the wind is almost as likely to blow the ball into the hole as it is to cause it to miss the hole. At 3 miles per hour the deflection is only half of the typical standard deviation of 10-foot putts as measured with several players, but still provides a plausible excuse for missing the putt. Above 4 miles per hour the effect is serious and players should be making careful allowances for the effect of the wind.

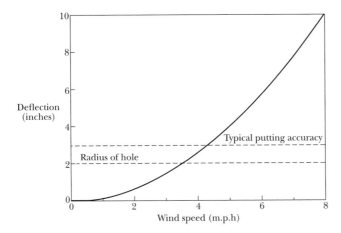

Fig. 9.23. Dependence of the ball's deflection on the ground wind speed for a 10-foot putt.

Biased balls

The balls, of course, are not perfect. The material of which the balls are made is not perfectly uniform and the shape of a ball cannot be exactly symmetrical. This means that all balls have a bias and, no matter how small the bias, a putt on a flat and level green will take a slightly curved path. The practical question is whether the bias has a significant effect.

The simplest way of regarding the bias is to think of the centre of gravity of the ball being to one side of the geometric centre of the ball. The centre of gravity will be in a particular direction from the centre of the ball and the half of the ball in that direction will be slightly heavier. In order to investigate the actual behaviour of balls it is necessary to be able to identify the direction of their bias. How can this be done?

If golf balls were less dense than water they could be floated and the lighter side of the ball would rise to the top. In fact the density of golf balls is approximately 1.12 times that of water. However, if salt is dissolved in water its density is increased. The water in the Dead Sea has a density that is 1.16 times that of normal water. It is therefore quite straightforward to dissolve salt in water to produce a solution in which golf balls will float. When a ball is so floated it turns until the lighter side is at the top and this side can then be marked with a pen.

In order to find whether the bias of balls could be detected in practice, five balls of different makes were floated in salt-water and their lighter sides marked. A golfer with a good putting ability then played all five balls from a position 30 feet from the hole. Each ball was played seven times with the bias on one side and seven times with it on the other. The final position of the ball for each putt was marked and its distance from a straight line through the centre of the hole was measured and recorded.

There will obviously be a spread in the final positions resulting just from chance, and the question is whether the effect of the bias can be separated from this spread. When the heavier side of the ball is placed on the left we expect the ball to move to the left and when on the right the ball would move to the right. Figure 9.24 shows the outcome of the experiment, the average final position of the right biased balls being 7.3 inches to the right of the left biased balls. A statistical analysis shows that the odds against a separation as large as this occurring by chance are more than a thousand to one.

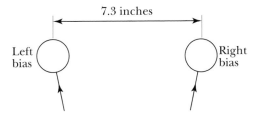

Fig. 9.24. Showing the average effect of the inherent bias of the balls on their final position in 30-foot putts.

The result means that the bias led to more than 3 inches error in each direction. Typically, of course, balls will not sit at an angle to give the maximum bias and so, on average, the effect for these 30-foot putts would be less than 3 inches. However, since the radius of the hole is approximately 2 inches the bias can clearly have an important effect for long putts. An ameliorating factor is that for most players the inaccuracy of their putts would typically be three or four times larger than the effect of the bias.

Muddy balls

In addition to the intrinsic bias of the balls there can be the bias that results from mud attached to the ball. On the green, players will normally clean a dirty ball but from off the green a putt must be taken with the mud in place.

To investigate the effect, putts were carried out with a small weight of 180 milligrams attached to the ball. This weight is about a quarter of the weight of a small drawing pin and about 0.4% of the weight of the ball. Ten putts were made with a left bias and ten with a right bias, all from 15 feet. Figure 9.25 shows the result. Eight of the right biased balls went to the right and nine of the left bias balls went to the left. The odds against as many as this going in the bias directions just by chance are over a thousand to one.

Fig. 9.25. Plot of the sideways displacement from the hole of 15-foot putts with the bias placed on the left and on the right.

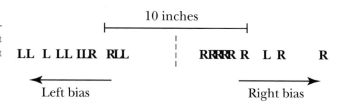

10 inches

LL L LL LR RLL　　　RRRR R L R　　R

←——————— Left bias

Right bias ———————→

The separation of the average final position of the two cases is 8.8 inches implying a 4.4-inch error introduced by the bias. If we take the error to be proportional to the added weight, this indicates that only 100 milligrams of mud could cause a straight 15-foot putt to miss the hole. Figure 9.26 gives an illustration of the size of a circular patch of mud of this weight.

Fig. 9.26. Illustration of the size of a mud patch that is just large enough for it to cause a straight 15-foot putt to miss the hole.

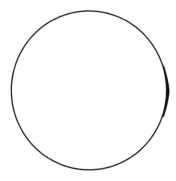

10 HOLE-IN-ONE

The history of golf is dotted with hole-in-one exploits. The first recorded case is that of (Young) Tom Morris who completed a 145-yard hole with one stroke in the 1869 Open Championship, which he subsequently won.

Holes-in-one are quite rare but nevertheless cases of successive holes-in-one have been recorded. Perhaps, the most remarkable is the performance of N. Manley who achieved successive par-4 holes-in-one, holing the seventh (330 yards) and eighth (290 yards) at the Del Valle Country Club in 1964.

It is difficult to assign a longest hole-in-one record because of the different conditions, with dog-legs and favourable slopes. The longest success on a straight hole is that by the American student Bob Mitera who holed his drive on a 447-yard hole—aided, however, by a downhill slope.

The most amazing hole-in-one achievement is that of the Leicestershire golfer Bob Taylor. In 1974 he holed the 188-yard sixteenth hole from the tee on the Hunstanton Links course on the practice day of a competition, and then produced the same feat on both the first two days of the competition proper—three successive holes-in-one on the same hole.

Probability of a hole-in-one

How often have you heard the question 'what are the odds for getting a hole-in-one?' The question is usually asked with the assumption that the answer is in the form of a single number, say 1000 to 1. There is, of course, no such single number.

The probability of a hole-in-one depends basically on two factors, the length of the hole and the ability of the player. There are other factors such as the protection of the hole by bunkers and the slope of the green but there is a limit to the amount of detail that is worth pursuing. In our analysis we will consider 'typical cases'.

The information we shall use to calculate the probability is rather diverse and comes from three sources. First, experiments were carried out to find the spread in the balls' landing positions together with their rolling distances, both for holes of different lengths. Second, a general view of probabilities can be obtained by discussing with individual players their hole-in-one successes or, more often, their lack of such success. Then, third, there are published estimates of the odds. These are usually quite unscientific but nevertheless do add some information.

Some interesting features arise. For example, the better players tend to use higher numbered clubs and the resulting back-spin gives the ball a shorter rolling distance with the associated reduction in the probability of finding the hole. This, of course, is offset by their greater ability to drop the ball near the hole in the first place. For longer length holes there is less difference in technique with all players needing to obtain sufficient range and finally resorting to woods and even drivers for longer holes.

With all of these aspects of the problem taken into account it is possible to estimate the dependency of the probability of a hole-in-one on the player's handicap and the length of the hole, and the results are given in the form of a contour plot in Figure 10.1.

Two comments are perhaps called for. First we should remember that the results represent a sort of average over holes of different types and over different conditions of play, and second, the estimated odds really represent a first attempt and it is to be hoped that they will be refined by future research.

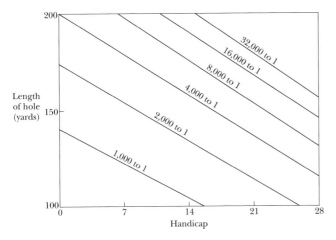

Fig. 10.1. The dependence of the odds against a hole-in-one on the length of the hole and the ability of the player.

How many?

We can carry out a rough calculation to estimate how many holes-in-one have been obtained. Perhaps there have been 40 million regular golfers worldwide. If we take the average number of rounds they have played to be 400, with 4 par-3s per round, then that makes a total of about 64 billion opportunities for a hole-in-one. If we take the average odds to be, say, 10,000 to 1, this gives an estimate of 64 billion/10,000 = 6.4 million holes-in-one. This is only a rough calculation but it seems quite likely that there have been more than a million successful cases.

11 HANDICAPS

The use of handicaps is the glory of golf. Imagine three friends playing a round of golf together having abilities corresponding to handicaps of 5, 15, and 25 but competing without the use of handicaps. There would be about a 95% chance that the best player would win with the weakest player coming last. A good handicap system would give them a roughly equal chance, completely transforming the game.

The origins of handicapping in golf lie in the early enthusiasm for betting on the game. In 1687 the Scotsman Thomas Kincaid was commenting in his diary on the relative merits of choosing reasonable odds and conceding strokes to even-up the chances.

The word 'handicap' has its seventeenth century origins in gambling, first appearing in a game in which, to quote the *Oxford English Dictionary*, 'a challenger laid claim to an article belonging to another person, offering something in exchange for it, the difference in the value of the two items being decided by an umpire who stood to gain the forfeit-money deposited by all three contestants if the other two parties both signified (by drawing out full or empty hands from a cap) their acceptance or rejection of his award (otherwise the one who accepted it won the forfeit-money).'

In the eighteenth century the term handicap was applied to a race between two horses. An umpire decided on a weight disadvantage to be imposed on a superior horse and the owners would signify their assent to or dissent from his adjudication, again by the way in which they withdrew their hands from a cap or hat.

The name handicap was introduced into golf in the nineteenth century, handicaps initially being used as an aid to betting, allowing some players to make a good income from the game.

English clubs adopted the use of handicaps in the 1870s. There is a report from 1875 of a 'handicap tournament' at which the players decided the handicaps to be used by 'a vote of all present'. It is hard to imagine players accepting such a procedure nowadays, although it might be fun to try it. A more generally used method was based on averaging the best three scores over the previous 2 years.

Initially, individual clubs allocated handicaps to their members. But between the years 1860 and 1900 the number of clubs in the British Isles increased from 36 to 1500, leading to increasing opportunities for players from different clubs to compete with each other and an obvious need for a national handicap system, providing 'portable' handicaps. The progress to a national system was rather slow but, with the Ladies Golf Union playing a leading part in the developments, a national system was finally achieved in 1926.

In 1983 a more sophisticated system based on that used by the Australian Golf Union was introduced. The governing body for the new rules was the Council of National Golf Unions (CONGU) comprising representatives of the Golf Unions of England, Ireland, Scotland and Wales, and the Royal and Ancient Club of St Andrews. The rules are often referred to as the CONGU rules.

In 2004 the English Women's Golf Association, the Scottish Ladies' Golfing Association, the Irish Ladies' Golf Union, and the Ladies' Golf Union joined with CONGU to create the 'Unified Handicapping System' to be implemented in 2008.

Until the twentieth century, handicaps in the United States were determined by first tee negotiation. The American handicap system has its origins in 1905, when the Metropolitan Golf Association adopted the British procedure of averaging the three best scores. In 1911 the United States Golf Association (USGA) introduced their first handicap system, a modified form of the British system. Over the following years there were further developments, often very controversial and unsatisfactory. The present system of taking the best 10 of the last 20 rounds was introduced in 1967.

The basic idea behind handicap systems is that if the players' handicaps are subtracted from their actual scores to obtain a Net Score, then the Net Scores level up the chances of each player winning. The idea is splendidly simple but the reality is more complex as can be witnessed from the clubhouse chat after a match or competition has been played. When players are asked what they believe the aim of their handicap system to be, they invariably say that it is to give all players an equal chance of winning. We shall find that the major handicap systems do not have this aim and do not provide equal chances.

We shall now examine the two major handicap systems, the British system, which involves very complicated rules to decide the handicaps, and the American system, which is basically quite simple.

THE BRITISH SYSTEM

The manual of the CONGU Unified Handicapping System runs to almost 100 pages and the account given below is necessarily a précis of the rules. It would not be surprising if readers find the CONGU handicap procedure mysteriously complicated, and some may prefer to skip the description of the system and move directly to the analysis of its consequences.

Par and Scratch Score

The par for each hole is decided by the club, and the overall par for the course is the sum of the pars of the individual holes.

However, the handicap system uses the Scratch Score. This is the score which a scratch player is expected to return. The Standard Scratch Score is allocated to a course by the Golf Union. In competitions an alternative Scratch Score, the Competition Scratch Score, is calculated according to specified rules.

Categories

The allowed handicaps have a range from 0 to 28 for men and 0 to 36 for women, and are divided into five categories:

Category	Handicaps
1	0–5
2	6–12
3	13–20
4	21–28
5	29–36

Net Score

A player's Net Score is his[1] actual score less his handicap. For example, a player with a handicap of 14 scoring 90 would have a Net Score of 90 – 14 = 76.

[1] The word 'his' will be used throughout to avoid the tedious repetition of 'his or her'.

Net Differentials

A player's performance is measured by his Net Differential, which is the difference between his Net Score and the Scratch Score. For example, a player having a Net Score of 76 with a Scratch Score of 72 would have a Net Differential of +4. If his Net Score were 70 the Net Differential would be –2.

The relationship of the Net Score and Net Differential is illustrated in Figure 11.1.

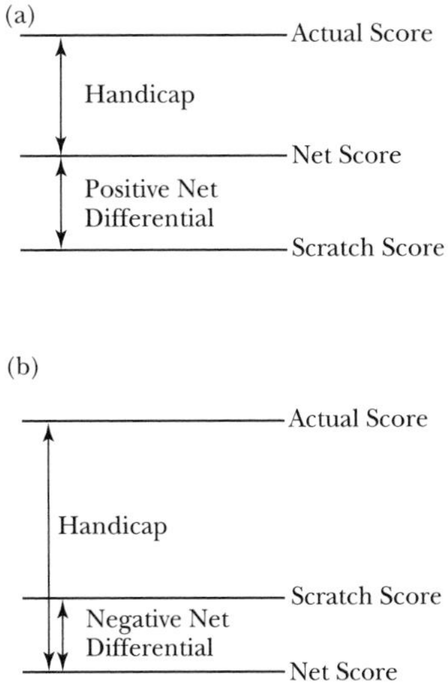

Buffer Zone

A score is within a player's Buffer Zone when his Net Differential is positive and lies within the following bands:

Category	Buffer Zone
1	0–1
2	0–2
3	0–3
4	0–4
5	0–5

Fig. 11.1. The figure shows how the Net Differential is determined from the actual score and the handicap for cases giving (a) a positive Net Differential and (b) a negative Net Differential.

(a)

(b)

Exact and Playing Handicaps

Players will have an Exact Handicap, which contains a decimal point, and a Playing Handicap, which is the nearest whole number. For example, a player with an Exact Handicap of 12.2 will have a Playing Handicap of 12. If the Exact Handicap involves a 0.5 it is rounded up to give his Playing Handicap.

Handicap Changes

If a player's Net Score is greater than the Scratch Score and the Net Differential is outside his Buffer Zone his handicap is increased by 0.1. For example, if a player with a handicap of 14.2 has a Net Score that is four strokes above the Scratch Score, this is outside his Category 3 Buffer Zone of 0–3 and his handicap will be increased by 0.1 to 14.3.

If a player's Net Score is greater than the Scratch Score but his Net Differential falls within his Buffer Zone his handicap is not changed.

If a player's Net Score is less than the Scratch Score, his handicap is decreased according to the table below. This gives the reduction of the handicap for each stroke by which the player's Net Score falls below the Scratch Score.

Category	Reduction for each stroke below the Scratch Score
1	0.1
2	0.2
3	0.3
4	0.4
5	0.5

For example, if a Category 4 player has a Net Score which is three strokes below the Scratch Score, his handicap will be reduced by $3 \times 0.4 = 1.2$.

The procedure of determining handicap changes is summarized in Figure 11.2.

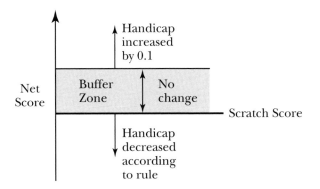

Fig. 11.2. Summary of procedure for making handicap changes under the CONGU handicap system.

Calculation of player's average Net Score

Having described the CONGU handicap system, we now turn to an analysis of its consequences. The first step is to address the theoretical task of determining the relationship between players' handicaps and their average scores. There are two ways of approaching the problem.

In the first approach we consider a player with a given ability and resulting average score. He will obviously obtain a spread of scores arising from the chance elements of the game – bunkers, mis-hits, and so on – and this will produce small modifications of his handicap, up and down. On average these will cancel out, the upward and downward changes balancing each other. Thus, allowing for the spread in the player's results, we can determine what average handicap will result from the balance between upward and downward changes.

This way we find the relationship between handicaps and average scores.

Although this procedure is theoretically sound it might seem strange to imagine the long time necessary for a given player's handicap changes to balance out. The second way of viewing the problem is to consider a large number of players on a given handicap and to average over their results to obtain the same balance between upward and downward changes. This gives the same result as the first method.

To carry out the calculation it is necessary to include the spread of the players' results. The bell-shaped normal distribution was introduced in Chapter 9 and we shall assume the same model here to describe the distribution of the scores. The spread of the scores is measured by their standard deviation and, from the analysis of players' results, typical values are found to lie in the range of 3–4.5 shots. With this information it is now possible to model the increases and decreases in handicap for players of given ability and determine the relationship between their handicap and their average score.

We initially take the standard deviation in players' scores to be 4.5 shots and calculate the difference between the average Net Score and the Scratch Score over the whole range of handicaps. The results are shown in Figure 11.3.

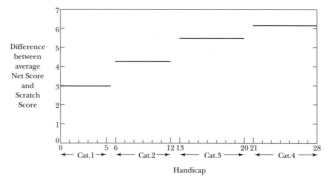

Fig. 11.3. The figure shows how average Net Scores vary with handicaps, assuming a uniform standard deviation of 4.5 shots.

It might be expected that the handicap system would arrange that players with different handicaps would all have an average Net Score approximately equal to the Scratch Score, or at least that they would have almost equal differences from the Scratch Score. However, we see that the better players in Category 1 have an advantage of 3 shots over the less skilled players in Category 4.

A further factor that influences the handicaps is the difference in the spread of scores for players with different handicaps. It is found that better players tend to have a smaller spread in their scores. Whereas Category 4 players typically have a standard deviation of 4.5 shots, Category 1 players have a standard deviation closer to 3 shots. In order to see the effect of this, the handicaps over the whole range were recalculated assuming a continuous variation of standard deviations from an average of 3 for the Category 1 players to average of 4.5 for Category 4 players. The results are shown in Figure 11.4. It is seen that the allowance for different standard deviations increases the calculated difference in scores for players with different handicaps.

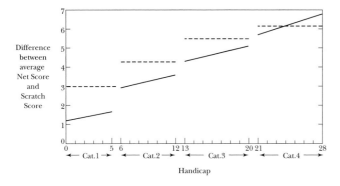

Fig. 11.4. Dependence of average Net Scores on handicaps, allowing for the differences in the spread of players' scores.

And in practice?

In the above analysis the precise consequences of the CONGU handicap system were calculated. In the real world things are not so tidy – we do not have access to the average score of the large number of players on each and every handicap value that would be needed for a scientific comparison. Nevertheless, the main feature of the CONGU system, namely the advantage given to low handicap players, can be examined by analysing some actual results. The scores of 36 players, with handicaps between 12 and 24, playing a combined total of 253 rounds, were used to calculate the dependence of the players' Net Scores on their handicaps. The resulting graph, shown in Figure 11.5, illustrates quite clearly the increase in average Net Scores with increased handicaps, consistent with the calculated results.

Fig. 11.5. Dependence of players' Net Scores on their handicaps.

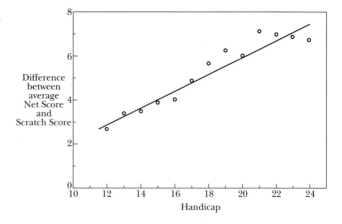

Simulations

To further illustrate how the CONGU handicap system
works we can simulate the changes in a player's handicap on a computer. A 'player' is allowed to play matches
in which his scores have his average score with a spread
given by his standard deviation. After each round his
handicap is adjusted according to the handicaps rules.
As an example we follow the handicap changes for
three players, all having an average score of 100 on a
course with a Scratch Score of 72. To make their average Net Scores equal to the Scratch Score they would
need to have a handicap of 100 − 72 = 28. So we start
off our players' rounds of golf with this 28 handicap
and calculate the subsequent handicap changes. The
results are shown in Figure 11.6.

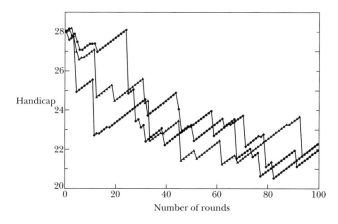

Fig. **11.6.** Simulated handicap
changes for players with an average
score of 100, starting with a handicap
of 28.

It is seen that the handicap system reduces all the handicaps to a lower level, after which they have random
variations about a value substantially lower than 28.
The sequence of handicap changes varies from simulation to simulation but the outcome is always the same.
Players who need a handicap of 28 to obtain an average
Net Score equal to the Scratch Score typically have a
handicap of 22.

The calculations show that the handicap system gives low handicap players an advantage over higher handicap players. A scratch player is given an advantage of about 5 shots over a 28-handicap player and a handicap-6 player has a 2-shot advantage over a handicap-20 player.

We have to conclude that, despite its complexity, the CONGU system does not provide players of different abilities with equal chances. The implications of the system for players' chances of winning matches and competitions will be analysed in Chapter 12.

There is another feature of handicap systems that is of interest to players – the time it takes for handicaps to respond to a change in a player's ability. For example, some players new to the game rapidly become good players and ideally the system would respond to that. Another situation that calls for a rapid handicap adjustment is when a player is debilitated by illness. The rules allow the player's Handicap Committee to make an ad hoc adjustment to players' handicaps but this is only necessary if the system itself does not make a sufficiently prompt correction. To illustrate the way the CONGU system responds, we take two players of equal ability, one of whom has a handicap 3 shots higher than his average score would imply and the other having a handicap 3 shots lower. We then use the computer to follow the subsequent changes in their Exact Handicaps. Figure 11.7 gives the results for typical cases where the players start with handicaps of 20 and 14. Both players move to their appropriate handicap of 17 with the inevitable ups and downs. However, in both cases the adjustment takes many rounds, the player with too high a handicap requiring about 20 rounds and the player with too low a handicap taking over 60 rounds. The obvious problem in this latter case is the slowness of the adjustment imposed by the limiting change of only 0.1 shot per round.

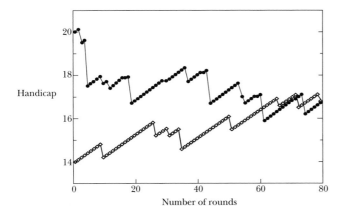

Fig. 11.7. The number of rounds required for players' handicaps to adjust is illustrated.

THE AMERICAN SYSTEM

The handicap system of the USGA assesses a player's ability by the attractively simple procedure of taking an average of the 10 best scores from his last 20 rounds. However, the actual calculation of handicaps involves a number of adjustments and we shall consider them in turn.

Equitable Stroke Control

It is thought that the score for a round can be distorted if there is a very high score on one hole, a score of 11 on a par-4 hole, for example. The handicap system corrects for this eventuality by limiting the allowed score on a hole. The maximum score depends on the player's handicap according to the following table.

Handicaps	Maximum scores
9 or less	2 over par
10 to 19	7
20 to 29	8
30 to 39	9
40 or more	10

This procedure is given the name of Equitable Stroke Control. Its significance is probably reduced by the fact that the 'unfortunate' scores are more likely to arise in the worst 10 rather than the best 10 rounds.

Course Rating

The difficulty of the course on which a score is obtained is measured by the USGA Course Rating, which is the expected average score of the best 50% of rounds played by scratch golfers.

The Differential

The Differential is the difference between the player's score and the Course Rating of the course on which he played. For example, if a player takes 100 strokes and the Course Rating is 72 then his Differential is 100 − 72 = 28.

Slope Rating

As we saw above, the Course Rating measures the difficulty of a course for scratch players. However, the difficulty of a course for players with higher handicaps can be quite different. We can, for example, think of a lake whose width is a challenge to a high handicap player but which presents no problem for a long driving low handicap player.

To allow for this factor, courses are individually rated to estimate the amount by which the difficulty of the course increases with the handicap of the player, or more precisely with the expected Differential of the player. The resulting rating is called the Slope. If we think of a graph in which the course difficulty is plotted against handicap, then the Slope Rating is a measure of the slope of this graph.

The Slope Rating for a course of standard difficulty is taken to be 113.

Handicap Differential

The Handicap Differential for a round is calculated using the formula

$$\text{Handicap Differential} = (\text{Score} - \text{Course Rating}) \times \frac{113}{\text{Slope Rating}},$$

rounding the number to the nearest tenth.

So, on a course having the average Slope Rating of 113, the Handicap Differential would be simply the number of shots by which the player's score exceeds the Course Rating.

Handicap Index

The Handicap Index is the player's personal handicap, which he carries with him from course to course. It is given by the average of his 10 lowest Handicap Differentials from the last 20 rounds. Well, not quite. This number is, curiously, to be multiplied by 0.96 so that,

Handicap Index = Average of 10 best Handicap Differentials × 0.96,

deleting all digits after the tenths.

Why 0.96?

The USGA gives the 0.96 factor the rather splendid name 'Bonus for Excellence'. The idea behind the 0.96 is that if the handicap system were to deliver perfect equality, all players having an equal chance, it would provide little incentive for a player to lower his handicap. The 0.96 disadvantages higher handicap players and gives a slight benefit for a reduced handicap.

In fact, as we shall see later, even without the 0.96 the USGA handicap system does not deliver perfect equity, it generally gives the lower handicap player an advantage. The 0.96 just increases that advantage.

Course Handicap

We have at last reached the bottom line. The handicap which a player will play off at a given course with its particular Slope Rating is

$$\text{Handicap} = \text{Handicap Index} \times \frac{\text{Slope Rating}}{113}.$$

The result being rounded to the nearest whole number.

An example

The table below gives an example of 20 rounds of golf, simulated on a computer, for a player playing on a course with a
Course Rating of 72 and the standard Slope Rating of 113.
The 10 best Handicap Differentials then correspond to the 10
lowest scores. The player's average score was taken to be 100
with a standard deviation of 4 shots.

	Score	Ten lowest scores	
	98	98	
	103		
	97	97	
	95	95	
	98	98	
	96	96	
	100		
	101		
	97	97	
	102		
	101		
	96	96	
	101		
	102		**Average of 10 lowest scores = 966/10**
			= 96.6
	93	93	
	105		**Handicap Index = (96.6 – Course**
			Rating) × 0.96
	105		**= (96.6 – 72) × 0.96**
			= 23.616
	98	98	**which becomes 23.6**
	110		
	98	98	
Total	**1996**	**966**	
Average	**99.8**	**96.6**	

The average of the player's 10 best scores turns out to be 96.6 and this gives a Handicap Index of 23.6. So, playing on this course the player would play off a handicap of 24. This would give him an average Net Score of $100 - 24 = 76$, 4 shots above the Course Rating. That the average score is in excess of the Course Rating is not a statistical accident, it is a consequence of the handicap system. It is clearly of interest to know how this excess score varies over the range of players' handicaps since this determines the player's chances of winning.

How much inequality?

If all players had the same spread in their scores the only inequality between players with different handicaps would be that arising from the factor 0.96. However, players with higher handicaps generally have higher standard deviations of their scores and this affects their handicaps. Calculation shows that the average of a player's 10 best scores is expected to be lower than his average score by 80% of the standard deviation of his scores. This means that by using the average of a player's 10 best scores rather than the average score, his handicap is also reduced by 80% of his standard deviation. This is brought out in Figure 11.8, which gives a smoothed graph of the probability of a player obtaining scores around his average score, again taking the spread in scores to have a normal distribution.

In the simulated example considered above the player had a standard deviation of 4 shots. We would, therefore, expect the average of his 10 best scores to be lower than his average score by 80% of 4, that is $0.80 \times 4 = 3.2$. Since his average score was 100 this predicts a 10 best-score average of $100 - 3.2 = 96.8$, quite close to the simulated average of 96.6.

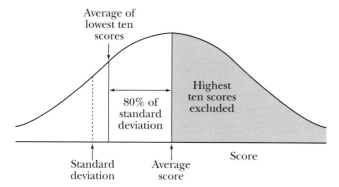

Average of
lowest ten
scores

80% of
standard
deviation

Highest
ten scores
excluded

Standard
deviation

Average
score

Score

Fig. 11.8. Distribution of scores about the average, showing how the average of the lowest 10 scores is related to the spread in a player's scores as measured by the standard deviation.

The average Net Score produced by the handicap system can be calculated for all players if we know their standard deviation and we assume that the Slope Rating makes the correct adjustment. The average Net Score will exceed the Course Rating by 0.8 of the standard deviation together with the effect of the Bonus for Excellence, which is $(1 - 0.96) \times$ Handicap. So we have,

Average Net Score – Course Rating =
$0.8 \times$ standard deviation $+ 0.04 \times$ Handicap.

It is found that the average standard deviation is about 3 shots for a scratch player and 4.5 shots for a handicap 26 player. Taking the standard deviation to vary smoothly between these values, and extrapolating to higher handicaps, we obtain a graph of the dependence of the difference of players' average Net Scores and the Course Rating, and this is shown in Figure 11.9.

Fig. 11.9. Handicap dependence of the difference between a player's average Net Score and the Course Rating.

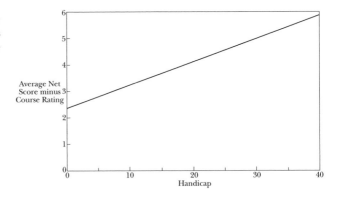

It is seen that a 24 handicap player has a 2-shot disadvantage when compared with a scratch player, roughly half coming from the difference in their standard deviations and half from the 0.96 factor.

If a player has an incorrect handicap, the USGA system steadily removes the deficiency, half of the error typically being corrected after 10 rounds and all influence of the incorrect handicap being removed after 20 rounds.

Comparison of British and American handicaps

The difference between the British and American handicap systems means that players of the same ability would have different handicaps under the two systems. If we take the American Course Rating and the British Scratch Score to be equivalent, and combine Figures 11.4 and 11.9 in Figure 11.10, it is seen that British scratch players would have a 1-shot advantage over their American counterparts whereas at a handicap of 28 the British players would have a 2-shot disadvantage.

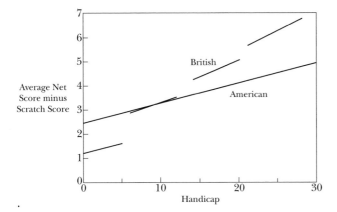

Fig. 11.10. Comparison of British and American handicaps.

Playing to your handicap

A misconception of some golfers is that they should be able to 'play their handicap', which implies that they expect a Net Score equal to the Scratch Score or the slope-corrected Course Rating. As we have seen in this chapter, the handicap systems are designed in such a way that players cannot expect to play to their handicap, and a player who regularly does so has acquired the wrong handicap.

So how often should a player play to his handicap? Using the results given in Figure 11.10 together with the assumed standard deviations it is straightforward to calculate the frequency of scoring the Scratch or slope-corrected Course Rating Score. The results are shown in Figure 11.11. It is seen that under the American system the frequency is about 1 in 6. Under the British system the frequency ranges from about 1 in 2 for scratch players up to 1 in 13 for players with a handicap of 28.

Fig. 11.11. The expected frequency with which players will 'play to their handicap'.

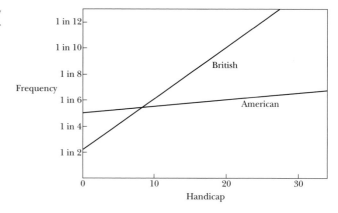

Wesson handicap system

Calculation of handicap changes under the CONGU handicap system involves a complicated procedure, and the USGA system requires a storing of the results from earlier rounds and a calculation for each player. For golf societies and groups of players playing regularly together there is a much simpler handicap system which gives the required handicap changes almost instantly. The system is simply this.

If the player's score differs from the median score of the group by more than 4, his handicap is changed by 1 shot.

The median score is the middle score, so if the players' cards are placed in the order of their scores, the middle card has the median score. If there is an even number of cards, the median is the average of the scores on the middle two cards.

If the competition is based on stroke-play, players with Net Scores more than 4 over the median have their handicaps increased by 1, if more than 4 under the median, then decreased by 1. Here is an example:

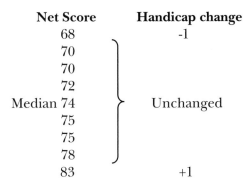

Net Score	Handicap change
68	-1
70	
70	
72	
Median 74	Unchanged
75	
75	
78	
83	+1

Stableford competitions will be described in the next chapter. In these competitions a higher score is a better score and so the handicap changes are reversed. Players with scores more than 4 over the median have their handicaps reduced by 1 and those with scores more than 4 lower than the median have their handicaps increased by 1. For example,

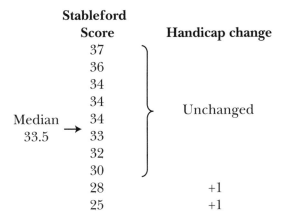

Stableford Score	Handicap change
37	
36	
34	
34	
Median 33.5 → 34	Unchanged
33	
32	
30	
28	+1
25	+1

This system works because it continuously moves players' scores towards the median score and the players towards an equal chance of winning. Over time, the movement of the scores towards the median is balanced by the chance variation in each player's scores giving a resultant spread of scores about the median.

There is no necessity to place a limit on the allowed handicaps. Newcomers should be given a provisional handicap, erring on the low side in preference to risking too generous a handicap.

The system can be modified for 9-hole competitions by taking more than three difference from the median as the requirement for a handicap change and making the change in handicap 2 shots. This produces an effective change of 1 shot over nine holes.

The operation of the system is illustrated by a real example. A group of players took part in a weekly nine-hole Stableford competition and their results were examined over a period of a year. The fortunes of two of the players show how the system works. One player's game improved during the year and the other's deteriorated. Figure 11.12(a) shows what their Stableford scores would have been if their handicaps had been unchanged. This displays the improvement in one player's score with an increase of 4 shots over the year, and the deterioration in the other player's score by 5 shots.

Fig. 11.12. Example of scoring of two players under the Wesson handicap system. (a) Stableford scores with no handicap adjustment and (b) actual scores with system operating.

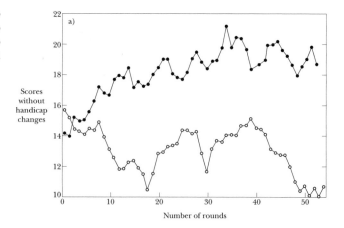

Scores without handicap changes

Number of rounds

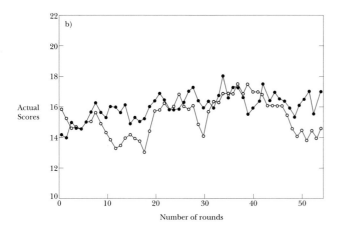

Number of rounds

Figure 11.12(b) gives a graph of the two players' actual scores, as modified by continual operation of the rule for handicap changes. It is seen that both players' scores are held quite steady. The two players' scores are also close to each other, the average difference between them being less than 1 shot, demonstrating the effectiveness of the system.

Although it is not essential, groups using this system can relate their handicaps to club handicaps by annually adjusting all of the handicaps by a chosen number of shots, the same number for all players. If some of the players in the group have club handicaps, the change should be such as to bring the average of their handicaps into line with the average of their club handicaps.

12 MATCHES & COMPETITIONS

Playing in matches and competitions is, in a sense, straightforward. You just concentrate and play the best you can. However, a theoretical description is very complicated. There are 18 different holes and the outcome at each one is determined by the difficulty of the hole and the ability of the player. Then each match or competition will involve different players each with their own handicap. Finally, and crucially, there is the element of chance—the ball rolls into a ditch, hits a tree, lands in a bunker, and drops, or does not drop, into the hole. Any attempt at a full description of all these factors leads to enormous complexity, and we shall avoid that by making simplifying assumptions and considering typical cases. Let us start by examining the expected outcomes for players with a scratch handicap playing average holes.

Scratch players

An examination of the scores of scratch players for average par-4 holes gave the results shown in Figure 12.1. This shows the probability, expressed as a percentage, of obtaining each score on the hole. It is seen that scratch players score par in about two-thirds of the cases and have an average score close to par.

If we take a golf course with 10 par-4 holes and a scratch score of 72, then an average score of 4.1 on the par-4 holes will contribute 41 to the player's overall score.

Figure 12.2 shows the distribution of scores on par-3 and par-5 holes. It is seen that scratch players lose shots on the par-3 holes with an average score of 3.4, whereas on par-5 holes they expect to score below par, almost half of their holes providing birdies.

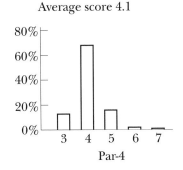

Fig. 12.1. Scores for scratch players on a typical par-4 hole.

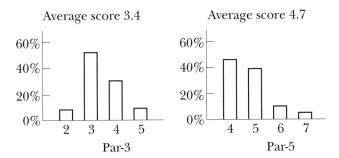

Fig. 12.2. Scores for scratch players on typical par-3 and par-5 holes.

If we take the course to have four par-3 and four par-5 holes, these holes contribute 4 × 3.4 = 13.6 and 4 × 4.7 = 18.8 to the total score. So, adding the scores for the par-3, par-4, and par-5 holes, we obtain a total average score of 13.6 + 41 + 18.8 = 73.4, slightly over par for the course.

Perhaps the most important aspect of the theory of matches is the question of how well players with different handicaps perform against each other. It is, therefore, useful here to make a comparison of the results for high handicap players with those of scratch players.

Higher handicap players

We take as an example players with a handicap of 27, and Figure 12.3 gives the comparison of their typical hole scores with those of scratch players as given above. The differences are clear in all the cases, but they are most obvious in the case of par-5 holes, where a 27-handicap player would be pleased with a bogey-6 but a scratch player would be slightly disappointed with a par.

Fig. 12.3. Comparison of the typical hole scores of a 27-handicap player with those of a scratch player.

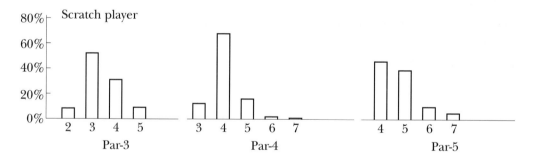

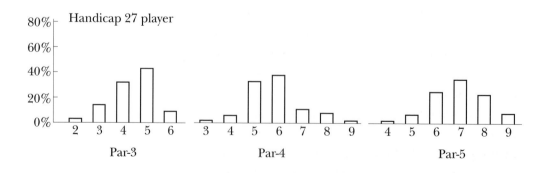

The need for simplification

If we consider the simplest match, two players playing match-play, there will be 20 or more possible pairs of scores at each hole, 3 and 4, 4 and 6, and so on. When we take 18 holes the total number of possible outcomes is astronomical. The description of a match could be simplified by considering only the result for each hole. There are then the three possibilities—player A wins the hole, player B wins, or the hole is halved. Over two holes there would by $3 \times 3 = 9$ possible outcomes, for three holes 27, and so on. Taking all 18 holes the total number of permutations is 387,420,489—each with its own probability.

This obviously calls for a simpler approach. What we shall do is to assume that a match will be won by the player who would obtain the lowest net score. This is not always the case of course but, on average, the errors will tend to cancel out.

Match played off scratch

In a scratch-match no handicap allowance is made. However, because both players have a spread in their expected scores, the weaker player will always have some chance of winning the match. As explained above, we shall estimate that chance by assuming the match to be won by the player who would obtain the lowest score over the 18 holes.

Each player will have an average expected score and the difference between the average expected scores of the two players will be given approximately by the difference in their handicaps. There will then be a spread of scores about the averages and for simplicity we shall take the two spreads to be equal and given by a typical standard deviation of 3.75 shots. As an illustration we take two players whose average scores are 80 and 83. Figure 12.4 then gives a graph of the probabilities of their scores, assuming normal distributions.

Fig. 12.4. Distribution of scores for two players whose average scores are 80 and 83.

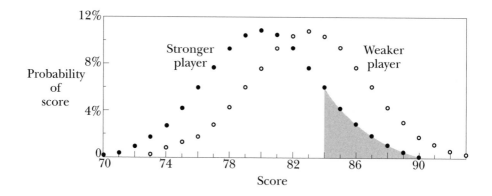

If, for example, the weaker player only gets his average score of 83, he will still win if the stronger player has a score of 84 or higher, the probability of which is about 17%, as indicated by the shaded region of the graph. There would also be an 8% chance of a half.

The probabilities of each player winning a scratch-match depend, of course, on the difference in the players' expected average scores. If the players have the same handicap, the probability of winning is 50% for both players. With a difference in handicaps the weaker player's chance diminishes as the difference between their handicaps increases. Making some simplifying assumptions, it is possible to calculate how the probability of a win depends on the difference in handicaps, and the result is given in Figure 12.5. With a handicap difference of 7, the weaker player's chance of winning is down to 10% and with double that difference his chance is essentially zero.

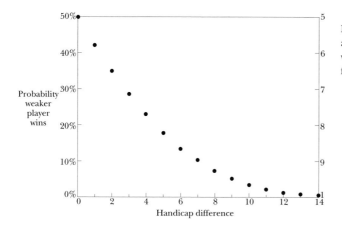

Probability—a digression

In the two previous chapters the concept of probability was introduced, relying on intuitive ideas of the subject. In Chapter 10 the probability of obtaining a hole-in-one was discussed in terms of the odds against and in Chapter 11 the spread in players' scores was described in terms of their standard deviation. In this chapter we shall use some elementary methods from the theory of probability, and it may be helpful to introduce these here before proceeding.

If a dice is thrown the probability of each number appearing is obviously one-in-six, that is, 1/6. So what is the probability of an even number appearing? We can either say that half the numbers are even and that the probability is therefore 1/2, or we can say that each of the numbers 2, 4, and 6 has a probability of 1/6 and so the probability of an even number appearing is 1/6 + 1/6 + 1/6 = 1/2, obviously the same answer. This illustrates the fact that the probability of one of several possible outcomes occurring is the sum of their individual probabilities.

If a coin is spun and a dice is thrown, what is the probability of a head and a six appearing? The probability of a head is 1/2 and the probability of a six is 1/6. The probability of them appearing together is therefore $1/2 \times 1/6 = 1/12$. This is obviously consistent with the fact that there is a total of $2 \times 6 = 12$ pairs of possible outcomes, each of which is equally likely. This illustrates that the probability of two independent events both occurring is given by multiplying their individual probabilities.

As an example of the application of these procedures to golf we can ask—if the probability of a player obtaining par on each of the first two holes is 0.4 and the probability of his obtaining a better score is 0.1, what is the probability that he will score par or better on both of the holes?

The probability of scoring par or better on a single hole is given by adding the probabilities, and this gives $0.4 + 0.1 = 0.5$. The probability of doing this on both of the holes is given by multiplying the individual probabilities, so the probability is $0.5 \times 0.5 = 0.25$.

Armed with these procedures we can now examine the probabilities involved in playing matches and competitions.

Match-play with handicaps—an example

If we consider a match between two players using their handicaps, we can say that, if we knew the probabilities of each score on every hole for both players, it would be possible to calculate the probability of winning for each player and of the match being tied. However, the required information is never available and in any case the calculation would be horrendously complicated.

What we can do is to illustrate the underlying principles with a simplified example. We will take a match between a scratch player and a player with a British handicap of 18. This means that the scratch player would give his opponent a shot on every hole. Taking a par-4 hole, the table below gives the typical probabilities of the two players taking each number of shots. The probabilities can also be expressed as percentages. For instance, the probability of 0.12 is the same as 12%. We note that for both players the sum of the probabilities is 1.00, that is, 100%, as it has to be.

Number of shots	3	4	5	6	7	8	9
Scratch player	0.12	0.69	0.16	0.02	0.01		
Handicap-18 player	0.01	0.23	0.34	0.25	0.09	0.06	0.02

We can now work out the implications of the table. If, for example, the scratch player takes 4 shots, he will win the hole if his opponent takes 6 or more. The probability of this is given by the sum of his probabilities for scores 6, 7, 8, and 9, and from the table this is $0.25 + 0.09 + 0.06 + 0.02 = 0.42$. Since the probability of the scratch player scoring 4 is 0.69, the probability of his scoring 4 and winning the hole is $0.69 \times 0.42 = 0.29$. On the other hand, with a score of 4 he will lose the hole if his opponent scores 4 or less, which has a probability $0.01 + 0.23 = 0.24$. So the probability that he will score 4 and lose the hole is $0.69 \times 0.24 = 0.17$. The hole would be halved if the scratch player scores 4 and his opponent scores 5, with a probability of $0.69 \times 0.34 = 0.23$. These calculations can be repeated for all of the scratch players' possible scores and the results are given in the following table:

Scratch Player's score	Probability that scratch player has that score and has a		
	Win	Loss	Half
3	0.09	0.00	0.03
4	0.29	0.17	0.23
5	0.03	0.09	0.04
6	0.00	0.02	0.00
7	0.00	0.01	0.00
	0.41	0.29	0.30

The final result is that the scratch player has a 41% chance of winning the par-4 hole and the 18-handicap player has a 29% chance, with the remaining 30% being the probability of the hole being halved.

The same calculation can be carried out using data for par-3 and par-5 holes. The results are strikingly different for the two cases. For par-3 holes the 18-handicap player is more likely to win, having a 44% chance compared to the scratch player's 27%. For par-5 holes the advantage lies with the scratch player who is almost twice as likely to win the hole as the 18-handicap player, with a probability of 51% as compared to 27%.

Handicap matches

When we examine matches played with handicaps we appreciate the importance of golf's handicap system. Figure 12.6 gives the calculated probabilities of winning for players under both the American and British handicap systems, depending on the difference in the players' handicaps. We see that whereas a player whose handicap is 12 greater than his opponent would only have about 1% chance of winning in a scratch-match, his chance is increased to over 40% in the American system and over 30% in the British. While there will be differences of opinion as to the merits of different handicap systems, Figure 12.6 is a dramatic demonstration of the crucial role of handicaps in making the amateur game so popular.

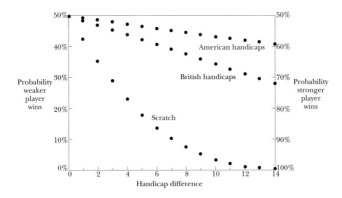

Fig. 12.6. Graphs showing the probability of players winning matches played with handicaps under the American and British handicap systems. The probabilities for a scratch-match are given for comparison.

Competitions

It is obvious that the probability of a player winning a particular competition depends on the number of entrants, the more entrants the less the chance. But how does this dependence arise? For simplicity we take a competition in which the handicaps give all players an equal chance. We can then calculate the probability that some player, one or more, will obtain each score. Figure 12.7 gives a typical graph of this probability for a competition involving 10 players, taking the standard deviation of their expected scores to be 4 shots. The highest probability, 65%, is for the average score, with a spread of about this value. The figure also gives a plot of the probability of each score being the winning score. It is seen that the most probable winning score is 6 shots below the average score.

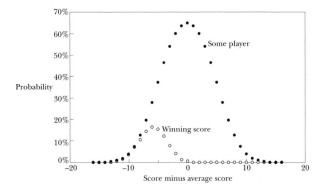

Fig. 12.7. The probability that, with 10 players, some player will score each score, together with the probability that the score is the winning score.

Figure 12.8 shows the corresponding graph for a competition with 100 players. This case has somewhat different character in that over a wide range of scores around the average it is almost certain that some player will have that score – the probability for the average score being over 99%. Understandably, the most probable winning score is now lower at 10 shots below the average.

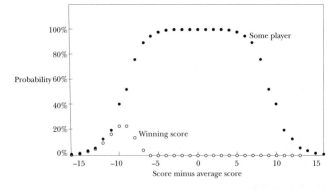

Fig. 12.8. The probability that, with 100 players, some player will score each score, together with the probability that the score is the winning score.

It is possible to calculate the most probable winning score for handicap competitions with any number of players. Figure 12.9 shows how, for typical competitions, the difference between the most probable winning score and the average score increases as the number of players in the competition is increased. It is seen from the graph that the calculation introduces fractional shot differences and it is reasonable to take the nearest whole number for each case.

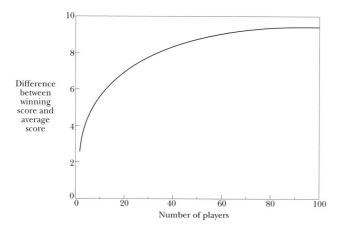

Fig. 12.9. Graph giving the dependence of the most probable winning score on the number of players in the competition.

Handicap bias in competitions

In the previous section the calculations were simplified by assuming that all the players in the competitions had the same average net scores and the same spread in their scores as measured by their standard deviation. We will now examine the effect of differences in players' handicaps on their chance of winning competitions.

As we saw earlier, differences in handicaps actually imply differences both in the players' average net scores and in the standard deviations of their scores. In competitions involving many players the effects of the two differences act in opposite directions. Higher handicap players will expect to obtain a worse, that is higher, score, but the greater spread in their scores gives them the opportunity of an occasional very low score.

The first point to make is that under the British handicap system the higher handicap player never has an advantage over lower handicap players. The difference between the higher handicap player's average score and the scratch score shown in Figure 11.10, is too large for the higher standard deviation of his scores to come to the rescue.

However, Figure 11.10 also shows that the American system of handicaps disadvantages higher handicap players less than the British system. As a result, it is possible under the American system for players with higher handicaps to have an advantage in competitions that have a sufficiently large number of players. In general, the theory of this effect is quite complicated but the essential features can be brought out by considering an example.

We will consider the case of two players with American handicaps of 5 and 25. Figure 12.10 shows the probability that their net scores will exceed the course rating, corrected for the slope, by a given number of shots.

Fig. 12.10. Probability of net scores for players with American handicaps 5 and 25.

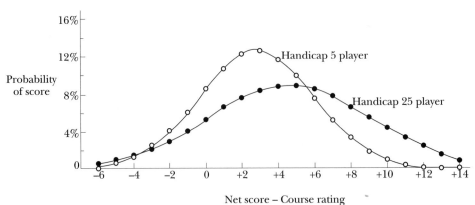

The most obvious difference is that the high handicap player has a smaller chance of obtaining a low score than the low handicap player and a greater chance of getting a high score. However, there is a subtlety. We see that the high handicap player is in fact more likely than the low handicap player to obtain net scores that are more than 3 shots lower than the course rating.

Let us now imagine that these two players play in a competition in which the competition average score is between their average scores with, say, an average net score of 4 shots above the course rating.

We can now use Figure 12.10 to assess their relative chances of winning the competition. The relevant part of the graph of Figure 12.10 is shown again in Figure 12.11, enlarged to make the situation clear. It is seen that if the winning score for the competition is 8 or more shots below the competition average score, the higher handicap player has a better chance of winning than the low handicap player.

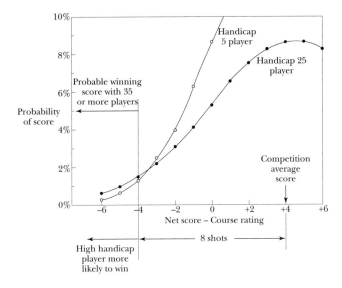

Fig. 12.11. The graph shows how, with American handicaps, it is possible for higher handicap players to have an advantage in competitions with a sufficient number of players.

We now need to know whether the winning score is likely to be 8 or more shots below the average score. This depends on the number of players and from Figure 12.9 we see that the critical number of players is 35. Thus we conclude that the 25-handicap player will have a better chance of winning than the 5-handicap player if there are more than 35 competitors.

The precise numbers in these calculations should not be taken too seriously because of the simplifying assumptions involved. However, the principle is clear—under the American handicap system a high handicap player can have an advantage over a lower handicap player if the competition has a sufficiently large number of competitors.

Stableford competitions

All golfers are familiar with the effect of a disastrous hole when playing in a stroke-play competition. A very high score on a hole usually removes the chance of winning the competition, and when it occurs at the start it can reduce the enjoyment of the rest of the round.

Dr Frank Stableford (1870–1959) was concerned with this difficulty, commenting that 'the thought ran through my mind that many players in competitions got very little fun since they tore up their cards after playing only a few holes', and Stableford, who was an excellent golfer, set about curing this problem by introducing a scoring system in which the effect of very bad holes was limited. His first suggested scoring system, introduced in 1898 at the Glamorgan Golf Club, was not entirely satisfactory and in 1932 he adjusted the system to its present form, the first Stableford competition being held at the Wallasey Golf Club in Cheshire.

Frank Stableford
(Wallasey Golf Club).

In Stableford competitions players are awarded points on each hole. The number of points depends on the player's Net Par for the hole. For a scratch player the Net Par is equal to the Par for that hole. A player with a higher handicap has the total Par for the course increased by an amount equal to his handicap. If the handicap is 18 or less, his Net Par for holes with Stroke Indices equal to or less than his handicap will be Par +1. If his handicap is above 18, his Net Par will be Par +2 for holes with Stroke Indices up to the amount by which his handicap exceeds 18 and will be Par +1 for the rest.

Two examples make it clear:

Handicap	Net Par
15	Par +1 on Stroke Indices 1 to 15
	Par on Stroke Indices 16 to 18
24	Par +2 on Stroke Indices 1 to 6
	Par +1 on Stroke Indices 7 to 18

Given the Net Par for each hole, players score according to the following table.

Score	Points
Net Par −2	4
Net Par −1	3
Net Par	2
Net Par +1	1
All higher scores	0

We see that Stableford arranged that there is no further loss of points for obtaining scores three or more higher than Net Par.

However, the Stableford scoring system is not quite what it seems. While it is comforting to players who play a bad hole and are not seriously punished, it will not usually affect the outcome of the competition. Generally, the player who wins a Stableford competition would have played well enough to avoid a bad hole.

We can investigate this by looking at stroke-play competitions. By examining the hole by hole scores it is possible to see how much benefit each player would have had if the scoring had been Stableford. Four stroke-play competitions were analysed, each with about 40 players whose average handicap was about 20. In none of the competitions would any of the top four players have benefited under Stableford scoring, and the winner would have won under either scoring system.

Fig. 12.12. Dependence of benefit from Stableford scoring on a player's net score.

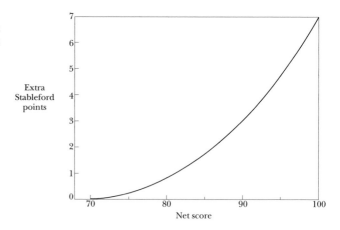

The situation is clearly illustrated in Figure 12.12 which combines the results from the four competitions examined to find the average benefit from Stableford scoring that would have accrued to players with each net score. The average net score of the winners was 69. It is seen from the graph that for all scores up to 8 shots higher than this, the average benefit from Stableford scoring is no greater than half a shot. The graph shows that the real advantage of the Stableford system is to players who have a bad round somewhat camouflaged by the system's generosity.

Nevertheless, Henry Longhurst could well have been right when he wrote of Frank Stableford, 'I doubt whether any single man did more to increase the pleasure of the more humble club golfer.'

13 THE
PLAYERS

It is clear that golf was being played in Scotland in the fifteenth century, because the Scottish King James II found it necessary to ensure the priority of military training by issuing the decree that 'golfe be utterly cryed downe and not be used'. Perhaps the military training was still not what it might have been—James II was killed by the bursting of one of his cannons at the siege of Roxburgh Castle.

Later, James IV renewed the ban more firmly with the statute 'in na place of the Realme there be used Fute-ball, Golfe, or uther sik unproffitable sportis' that were contrary to 'the common good of the Realme and defense thereof'. It is perhaps ironic that the earliest golfer of whom we have any record is just this same James IV, his purchase of golf equipment being listed in his accounts. James IV also came to a violent end, being killed on the battlefield at Flodden. However, the royal interest in golf persisted and James IV's granddaughter, Mary Queen of Scots, was famously enthusiastic about playing golf. It was regarded as in-criminating when she was seen playing golf only days after her husband Lord Darnley, from whom she was seriously alienated, met a mysterious death.

When Mary's son, James VI of Scotland, became King of England he took his golf clubs with him and the spread of the game was well underway. The rest of the Stuart kings all played golf but royal participation in the game died out with the accession of the Dutchman William III and his wife Mary in 1688.

The following two centuries saw an increasing popu-larity of the game. A report of the early days says 'The greatest and the wisest of the land were to be seen on the Links of Leith mingling freely with the humblest mechanics in pursuit of their common and beloved amusement'.

However, this state of affairs was to change with the formation of clubs. The initial purpose of clubs was to arrange golf competitions for 'Noblemen and Gentle-men', starting an exclusivity which persisted into the twentieth century and to some extent to the present day.

The first green at St. Andrews, from an engraving by Frank Paton dated 1798.
Source: Hobbs Golf Collection

The middle of the nineteenth century saw the introduction of championships. The forerunner of the present Amateur Championship was first held in 1857. The first 'Open' Championship took place in 1860, although for this first event the name Open is a misnomer since only professionals were allowed to enter. The following year the Championship was made truly open.

The credit for the organization of both the Amateur and the Open Championships goes to the Prestwick Club and the inaugural 1860 Open was held over three rounds of the 12-hole Prestwick links. Only eight players took part and their scores were:

Willie Park (Musselburgh)	174
Tom Morris (Prestwick)	176
Andrew Strath (St. Andrews)	180
Bob Andrew (Perth)	191
Daniel Brown (Blackheath)	192
Charlie Hunter (Prestwick St. Nicholas)	195
Alex Smith (Bruntsfield)	196
William Steel (Bruntsfield)	232

Tom Morris, who finished second, was a giant of the game. He was initially apprenticed as a ball-maker and at the age of 18 became greenkeeper at the Prestwick

Club. After the first Open he won four of the next seven Championships.

It is usual to refer to this Tom Morris as Old Tom Morris because his son, Young Tom Morris, was an equally remarkable golfer. In the early Open Championships the trophy was the Challenge Belt, and the Belt became the property of Young Tom Morris when he won it in the 3 years following his father's last win in 1867. At his first win Young Tom was only 17. Sadly he died at the early age of 24.

Tom Morris with Young Tom Morris, around 1870.

Prize money

In the first 3 years of the Open none of the players received a monetary prize. Even when prize money was introduced in the fourth year, only the runner-up and third and fourth players were rewarded, the winner being expected to be content with winning the Belt. However, in 1864 the winner received £6. This was equal to about 6 weeks of an average worker's wage, which makes it equivalent to about £3000 in today's money. Since those early days, the prize money has increased continuously as illustrated in Figure 13.1, which gives

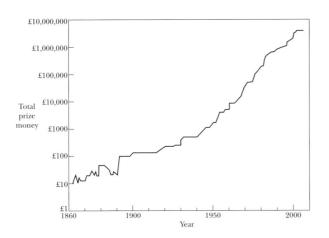

Fig. 13.1. Graph showing the increase in the total prize money at the Open Championship since its inception in 1860.

a graph of the total prize money each year at the Open since its inception.

The dramatic nature of the increase in prize money is brought out further by Figure 13.2, which displays the amount of the first prize in the Open in terms of the national average annual wage. Until 1950 the first prize was less than a typical person's annual wage, but since then the increase has been rapid and in the twenty-first century the first prize has risen to more than 25 times the national average wage.

Fig. 13.2. The first prize in the Open Championship in terms of the average annual wage.

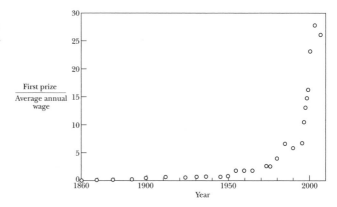

The American PGA Tour is the dominant element in the professional game, organizing a competition almost every week. The four most important of the professional competitions are the so-called Majors—the Open, the US Open, the Masters, and the US PGA Championship – and all of these competitions have a first prize of over a million dollars. Even for the less prestigious competitions on the PGA Tour most of the first prizes are around a million dollars.

In the Masters competition about 90 players take part, and in the other three majors there are about 150 players. The competitions involve four rounds played over 4 days, with about half of the entrants missing the 'cut' and dropping out after 2 days. The distribution of the prize money is illustrated by Figure 13.3, which gives a graph of the money won by players in the 2006 US Open plotted against their final place in the competition. Fifteen players received more than $100,000 and all 63 who made the cut received more than $15,000. Those who missed the cut were paid $2000.

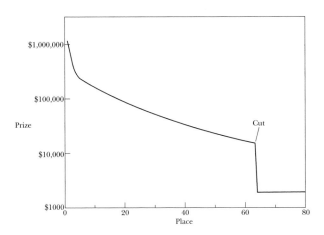

Fig. 13.3. The dependence of the prize money received by players at the 2006 US Open on their final place in the competition.

Distribution of ability

Can we assess the distribution of ability over the whole range of golfing abilities, from Tiger Woods through to players on the highest handicaps? This cannot be done precisely mainly because of the uncertainty about the abilities of casual golfers. However, if we restrict ourselves to regular players then we can use the distribution of the handicaps of club members as an indication for amateur golfers and the results from the professional tours for the top players. Combining these we shall be able to make a reasonable assessment of the 'universal' distribution of abilities.

Club players

There is, of course, some scatter in the distribution of handicaps within a club, just by chance there might, for example, be a lot of players on a particular handicap. However, we can represent the number of players with each handicap by a smooth curve that reasonably approximates the actual values. Taking a sample of British golf clubs it is found that the distribution of handicaps is well described by the bell-shaped normal distribution, and three typical examples are shown in Figure 13.4.

Fig. 13.4. Three examples of the distribution of handicaps for members of golf clubs.

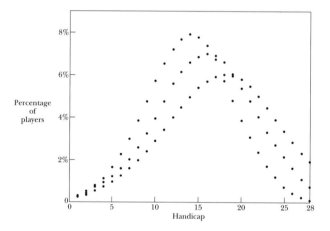

Generally the peak in the distribution occurs between handicaps of 14 and 20. It is seen that the curve falls off rapidly towards zero handicap. At the upper end of the distribution it is clear that in some clubs at least, there will be players whose natural handicap is above the British allowed upper limit.

Professional players

Most of the world's top players play on the US PGA Tour. The players' performance on this tour is analysed in great detail, with statistics covering every aspect of the game. Of particular interest are the scoring averages for each player, since these scores enable us to see the distribution of abilities. Figure 13.5 gives a graph of the average score against the rank of the player for the year 2006. The top player is Tiger Woods with an average score of 68.1. Fourteen players have average scores within 2 shots of this score and a hundred players are within 3 shots. These statistics bring out clearly the competitiveness of professional golf at the top level.

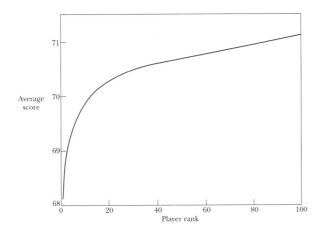

Fig. 13.5. The dependence of players' average US PGA Tour scores on the rank of the players for the year 2006.

Universal distribution

The results of the last two sections give us the opportunity of compiling the distribution of abilities of the world's golfers across the whole range of abilities. However, we should not take this too seriously because of the uncertainties involved. One of these is the number of golfers in the world. There are probably about 60 million people who play golf but most of these are casual golfers. We shall take the number of regular players to be 20 million. Using the data in Figures 13.4 and 13.5 we can then estimate the number of players who would, on average, obtain each score on a standard Par 72 course, and the result is shown in Figure 13.6. While lacking precision, the graph gives an indication of the spread in abilities and how good a player has to be to join the elite.

Fig. 13.6. The number of players worldwide expected to obtain each score on a typical course.

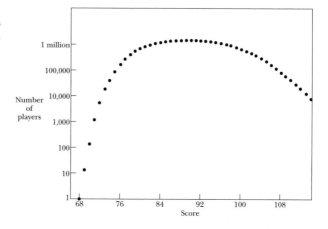

Ability and age

The youngest winner of one of the four Major Championships was Young Tom Morris, who was 17 when he won the Open in 1868. The oldest winner was Julius Boros who in 1968, exactly a century later, won the US PGA Championship at the age of 48. These are the extreme ages and between them there will be a distribution of the winners' ages. What is the most likely age for a winner in the Majors?

We can take the ages of Major winners who have completed their careers and find the ages at which they won their Championships. Figure 13.7 gives a graph of the number of wins in the four Majors for players who had six or more wins. This is plotted against their age at the beginning of the year in which their win took place. There is, of course, the usual scatter but the smooth curve represents the points fairly well, giving a most promising age of around 32 or 33. Half of the winners' ages were in the 7-year range between the ages of 29 and 36.

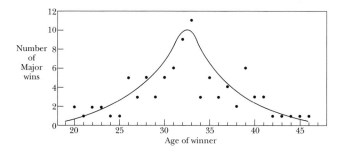

Fig. 13.7. The dependence of the number of wins at the Major Championships on the age of the winners.

Comparison with football (soccer)

It is interesting to compare the age profile for Major wins with a similar profile for goal scoring in football.[1] Figure 13.8 shows the annual scoring rate for top-level strikers and compares it to the corresponding curve for Major wins, as given in Figure 13.7.

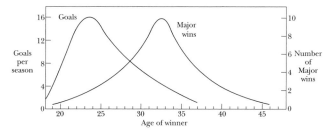

Fig. 13.8. Comparison of the age distribution of Major winners with the age dependence of the scoring rate for strikers in football.

[1] J. Wesson, *The Science of Soccer* (Institute of Physics, 2002).

The peak of scoring ability for footballers occurs between the ages of 23 and 24. So golfers reach their peak about 9 years later than strikers. This is presumably due to the longer period required to develop the complex array of skills required for golf as compared to the need for peak fitness and fast reactions which exemplify the best strikers.

Scores—what counts?

What skills are most important in obtaining a good score? There are of course many requirements, but perhaps the most often quoted are the ability to hit a long drive and skill on the putting green. To obtain some insight into the role of these factors the author carried out a series of experiments on the course.

The experiments used a typical par-4 hole with a length of 402 yards. The fairway is encumbered with the usual share of trees and bunkers together with a water hazard. Thirteen players took part and each one played the hole nine times, taking three balls from tee to hole three times. The total number of balls played was therefore $13 \times 9 = 117$, and for each case the score, driving distance, and number of putts on the green was recorded.

The players' handicaps ranged from scratch to the British upper limit of 28, and Figure 13.9 gives the basic result with a plot of average score against handicap. Of course, the graph has some scatter but the trend is clear. There is a variation of more than 2 shots over the range of handicaps, with the best players averaging par and the higher handicap players getting a double bogey. It might be expected that the range of scores for one hole would be the range of handicaps divided by the number of holes on the course, that is, $28/18 = 1.6$

shots, but the larger range of scores demonstrates the feature of the handicap system described in Chapter 11, namely that the system puts higher handicap players at a disadvantage.

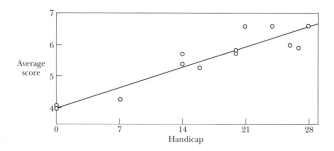

Fig. 13.9. Results of experiments giving the dependence of players' average par-4 hole scores on their handicaps.

What gives the low scorers their advantage? Figure 13.10 gives the dependence of players' scores on their driving distance. The trend line shows that the variation in average scores can be mainly attributed to the differences in the players' ability to hit the ball a long way.

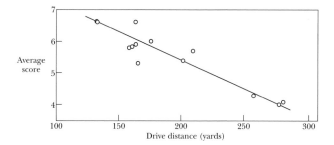

Fig. 13.10. The dependence of players' scores on the distance of their drives.

The other factor we need to look at is how the number of putts taken on the green varies over the range of handicaps. The results are shown in Figure 13.11, which plots the average number of putts on the green against the player's handicap.

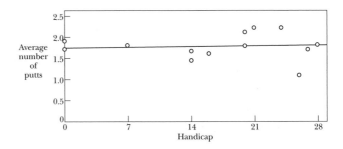

Fig. 13.11. The dependence of the number of putts on the green on players' handicaps.

There is again the expected scatter but the trend line shows that there is little difference in the average number of putts over the whole range of handicaps. It is reasonable to infer from this that the differences in putting ability play a smaller role than do the differences in driving distance.

However, the interpretation of the putting results is not entirely straightforward. The longer hitters will often be placing the ball on the green with their second shot, typically from a distance of 150 yards. From this distance it is difficult to place the ball near the pin and the result is a longer first putt. On the other hand, the short hitters will be approaching the green from shorter distances and sometimes using a putter from the fringe. This means that their first putt on the green may be somewhat shorter.

In order to clarify the role of putting ability a separate set of experiments was carried out to investigate the dependence of putting ability on the player's handicap. Each participant was asked to putt 60 times to a hole on a level putting green. The first five putts were from 1 foot, the second five from 2 feet, and so on up to a distance of 12 feet. The successes were recorded for each participant and the results were then analysed to determine the distance from which they would expect a 50% success rate.

Figure 13.12 gives a graph of the 50% distance plotted against the player's handicap. These results add support to the view that, although putts account for about a third of the total number of strokes, the differences of putting ability across the range of handicaps is small and this reduces significantly the role of putting. The strong dependence of scores on driving distance shows that hitting distance is the predominant factor in determining players' handicaps.

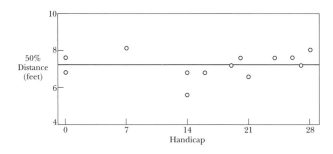

Fig. 13.12. Results of putting experiment, showing the distance for a 50% chance of a successful putt for players with a range of handicaps.

14 CLUBS & BALLS

Over the last two centuries the game of golf has been transformed by improvements in the equipment. These improvements have been brought about by better design, by advances in manufacturing techniques, and by the availability of new materials.

We have now reached the point where all manufacturers have a good understanding of the scientific principles involved and all have access to essentially the same techniques and materials. The result is that, although manufacturers are keen to draw attention to the advantages their equipment may offer, the differences are quite small. Of course, for professional golfers it is very important that they have the equipment best suited to their needs. For the ordinary golfer, buying a small improvement in their equipment means, in practice, that they might have the satisfaction of taking a shot or two off their handicap.

Early clubs

The earliest record of the manufacture of golf equipment comes from the Scottish Lord High Treasurer whose accounts for the year 1502 include the entry:

The xxi day of September, to the bowar (bowmaker) of Sanct Johnestown (Perth) for clubbs   xiiijs.

and at a later date:

For Golf Clubbis and Ballis to the King that he playit with   1xs.

The King referred to is James IV of Scotland. His enthusiasm for the game was passed to his descendants including his great-grandson James, who became the King of England and is recorded in 1603 as appointing William Mayne to be the royal fledger, bower, club-maker, and spear-maker for life. It looks as though the bow-makers had found a profitable side-line in supplying golf clubs.

None of these seventeenth century clubs have survived. The oldest clubs that exist are from the eighteenth century. These clubs have a shape which is closer to that of hockey sticks than that of modern clubs and they were also much longer. The golf historian Robert Browning remarks, 'They are so long in the shaft that it is difficult to see how the old heroes of the game contrived to hit the ball at all' and 'Apparently our great-grandfathers favoured what would now be considered an excessively flat swing, and stood very far from the ball'.

(a) An early nineteenth century long-nosed wood; (b) A wood made by Hugh Philp around 1850; and (c) A blacksmith's iron, around 1790. Source: Hobbs Golf Collection.

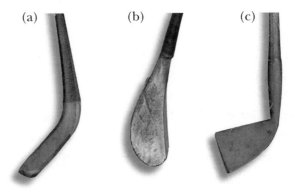

(a)   (b)   (c)

In the early times the approach shots were made with wooden clubs with a range of lofts, being called short spoons, mid-spoons, long spoons, and baffing spoons, the baffing spoon playing the role of a wedge. The use of iron for clubheads was unsatisfactory while the feathery balls were still in use, their leather cover being too easily cut by an iron blade. The introduction of the gutty ball opened the way for iron clubs. Another development was the introduction of the steel shaft which became a competitor for the conventional shaft made of hickory.

Modern irons

It is understandable that there was a reluctance to move from the wooden headed clubs to irons. The design and manufacture of wooden clubs had produced clubs which were a work of art. They were beautifully shaped and polished, with smart inserts to protect them from damage.

Golfers generally built up their collection of clubs on an ad hoc basis, picking up a mashie here and a niblick there. Along with the introduction of steel clubs came the idea of matched sets of clubs. This meant that, faced with their next shot, players had a choice from a set of similar clubs with a well-defined gradation of lofts. Golfers came to prefer this system and now irons are fairly well standardized, a typical set having lofts as given in Table 14.1. It is seen that the clubs are separated by a loft difference of 4°.

Iron	Loft
3	22°
4	26°
5	30°
6	34°
7	38°
8	42°
9	46°
Pitching Wedge	50°
Sand Wedge	54°

A further development in the design of irons was the introduction of perimeter weighting, in which there is a redistribution of the mass of the clubhead with a concentration towards the edge of the club as shown in Figure 14.1. The expected benefit from this is a reduction in the twist of the clubhead when the ball is hit off-centre. The distribution of mass towards the edge of the club increases the club's moment of inertia, the moment of inertia being a measure of a body's reluctance to rotate when subjected to a torque, in the same way that the mass of a body is a measure of its reluctance to accelerate when subjected to a force. Figure 14.2 illustrates a typical cross section of a perimeter-weighted club.

Fig. 14.1. A perimeter-weighted iron.

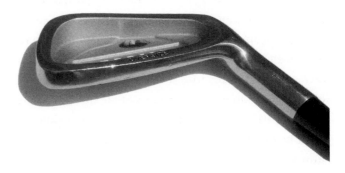

Fig. 14.2. Cross section of a typical perimeter-weighted iron.

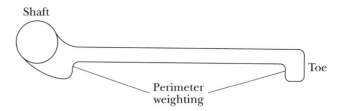

Modern drivers

Woods were little changed before 1980. The optimum design had been achieved by trial and error and skilled craftsman produced the clubheads from solid blocks of wood and inserted metal plates in the sole of the club to prevent wear through contact with the ground.

In the 1980s drivers with stainless steel heads became popular. Because steel has a density that is about eight times greater than that of hard woods it is clear that a steel clubhead of the same size and weight as a wooden clubhead would have to be hollow. The result was a club with an average wall thickness of about a sixteenth of an inch.

A wooden wood.

More recently, there has been a move towards substantially larger clubhead sizes. Many drivers now have a volume that is close to the maximum allowed by the rules, 460 cubic centimetres, which is 28 cubic inches. This, of course, has required the wall thickness to be reduced, roughly halving the thickness but with the face of the club thicker than that of the sole and crown. Generally the head shape and sole plate are cast separately and then welded together. The resulting clubhead is very durable, contributing to the popularity of iron woods.

The playing advantages of large steel heads are twofold. First there is a similar advantage to that of perimeter weighting in irons—the increased moment of inertia. Actually, a substantial increase had come with the redistribution of the mass when solid wooden clubheads were first replaced by thin-walled steel heads. The overall result is that the large steel clubhead has a moment of inertia which is more than double that of a wooden clubhead.

The second advantage of the large heads, presumably mainly for less skilled players, is the forgiveness of having a larger clubface with its reduction in the number of mis-hits, together with the resulting increased confidence in the swing.

Hybrids

A recent, and perhaps somewhat belated, development is the introduction of hybrid clubs, also called utility or rescue clubs. These clubs take on a role between that of the fairway wood and a low-number iron, making possible a more refined club selection. Figure 14.3 shows the cross section of a typical hybrid clubhead and compares it to those of an iron and a wood. This figure leaves the impression that there is a place for a further club between the iron and the hybrid.

Fig. 14.3. The relation of the geometry of a typical hybrid clubhead to that of an iron and a wood.

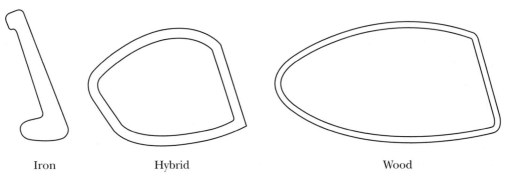

Iron   Hybrid   Wood

Putters

In earlier times putters had simple heads made of wood. When metal heads were introduced they generally had the form of a simple slab about half-an-inch thick. In more recent times heads with a marvellous variety of shapes have appeared. Since virtually all these shapes find owners who are quite satisfied with them, we have to conclude that putter performance is quite insensitive to the design, provided of course that the optimum strike point is clearly marked. It is interesting that most professional golfers use a simple putter.

A diversity of putter shaft lengths has also appeared and now we see a range of lengths from 34 to 60 inches. The long shaft requires a completely different technique, but some players who have mastered this come to prefer these shafts. However, it is rare to see amateur players using the very long shaft.

Shafts

The important characteristics of shafts are their flexibility and their weight. Shafts are graded according to stiffness by the manufacturers into five classes – L (Ladies), A (Amateur or Senior), R (Regular), S (Stiff), and X (Extra stiff). However, there is no agreed standardization and as a result there is some overlap between the classes. The methods of measuring stiffness by bending and vibrating the shaft were described in Chapter 2. The full range of stiffnesses as measured by the bending test gives a variation of about 30%, and this implies a range of vibration frequencies of about 15%.

It is perhaps surprising that there is little difference in the flexibility range of steel and graphite shafts. The main difference between the materials is their weight, with graphite shafts typically having half the weight of steel shafts. This difference means that graphite shafts give a slightly higher clubhead speed.

Balls

The early development of golf balls was described in Chapter 4. The main design changes were driven by the need for durability and the discovery that balls with a non-smooth surface flew better than their smooth counterparts.

Balls made of gutta-percha came into widespread use after the 1860s. Gutta-percha is a hard rubber-like material obtained from a species of trees mainly indigenous to Malaya. The 'guttie' balls had the advantage that they were only a quarter of the cost of the featheries they replaced, together with their ability to be re-moulded when worn.

Another revolution in golf ball manufacture was initiated in America when Coburn Haskell and Bertram Work introduced a rubber-cored ball. This ball consisted of rubber threads wound round a central core and encased in a gutta-percha shell. The large-scale production of the Haskell ball was made possible by the design by John Gammeter of a machine able to wind the threads.

By the early twentieth century it became clear that the 'Haskell' was superior to the guttie, allowing a substantial increase in the length of both drives and fairway shots. This development led to concern that the game was being altered in an uncontrolled way. As a result, the governing bodies in Britain and America separately attempted to place limits on the size and weight of the ball. This led to 70 years of controversy as the two sides tried to find agreement on the rules. By 1931 they had both settled on a maximum weight of 1.62 ounces. For the next 60 years British and American golfers used different sized balls, with a minimum diameter of 1.68 inches in America and 1.62 inches in Britain. The difference was finally resolved in 1990 when the Royal and Ancient adopted the 1.68-inch limit.

The changed character of the golf ball is brought out by the excerpt from an 1899 copy of Golf Illustrated, reproduced in Figure 14.4.

A Historic Incident.

WE are glad to be able to give pictures, from snap-shots taken by Mr. W. W. Macfarlane, of Edinburgh, of the sensational incident which occurred in the final round of the Amateur Championship at Prestwick. The scene is the bunker at the Alps hole, the 17th. The bunker was full of water and both Mr. Ball and Mr. Tait found it with their second shots—Mr. Tait's ball lying in the middle of the water and Mr. Ball's on sand just on the edge. With great heroism Mr. Tait waded in over his boots and successfully played the floating ball on to the putting green. Mr. Ball followed suit with an equally brilliant shot, he also having to stand in the water to play it.

MR. TAIT PLAYS HIS BALL ON TO THE GREEN.

Fig. 14.4. *Golf Illustrated*, 16 June 16 1899.

The material used in the construction of the balls has undergone continuous development. The gutta-percha casing was soon replaced by balata, a latex obtained from trees native to the north-eastern part of South America. Balata provided improved colour and toughness together with better adhesion to the rubber windings. From the 1940s, balata was itself replaced by synthetic plastics and in particular by Surlyn, a tough thermoplastic polymer developed by DuPont.

The next change came in the 1980s, when the development of synthetic materials reached the point where a quality two-piece ball could be manufactured. These balls have a uniform, hard rubber core covered with a thin, synthetic casing.

The two-piece ball now competes with a three-piece ball that has an inner core with a mantle layer inside the outer casing, some players preferring the softer feel of the three-piece ball and others the two-piece ball for its slightly greater distance.

Ball manufacture

The core of the two-piece ball is rubber, mixed with a blend of powdered fillers. This material is placed inside steel moulds and subjected to intense heat and pressure. The high temperature creates a chemical reaction that hardens the core as it cools. Excess material is then removed, leaving the required spherical shape. The cover is added in one of two ways – using either injection moulding or compression moulding.

With injection moulding the cores are placed in a mould base and held centred by small pins, leaving a narrow cavity around the core. The mould is then closed and the plastic cover material is injected into the spherical cavity. The mould casing has a raised dimple pattern which leaves its imprint on the ball. As the plastic cools and hardens the pins are retracted. In compression moulding the cover is first injection moulded into two hollow hemispheres. The two halves are then heated and pressed together around the core using a mould that imposes the dimples.

With the cover now in place, the balls are polished and painted before being given a logo and a final spray of clear paint.

Improved performance

It is often suggested that the improved performance arising from the technological improvements in clubs and balls is harming the game. It is thought that, particularly at the professional level, the challenge of the game has been lowered and that the strategic subtleties are being removed. On the other hand, it is argued that the improvement in performance cannot be attributed entirely to the equipment because allowance must be made for the increased fitness and skill of the players.

The actual improvement in performance is brought out most clearly in the increased length of the drives on the United States PGA tour. Each year there is a player with the longest average driving distance, and Figure 14.5 gives a graph of these distances over the last 25 years. It is seen that there has been a substantial increase in the drive length of about 45 yards. The average rate of increase over this period is about 6% for each 10 years, with a higher rate of increase of 10% over the last 10 years.

Can we assess how much of this improvement is due to the players themselves? Some idea of what to expect can be obtained by looking at the improved performances in other sports. Perhaps, the most relevant comparison is with the throwing events in athletics, where success is again measured by distance. In the shot-put and the hammer, discus and javelin throws, the equipment is standardized and therefore plays no role in any improved performance.

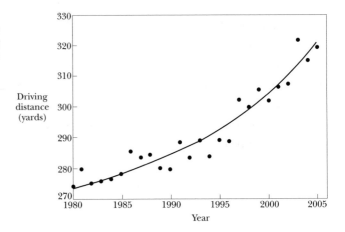

Fig. 14.5. Showing the increase in the annual average driving distance of the longest hitter on the US PGA tour.

We shall use the progression of the world record distances as a measure of the improved performance in these events. The world records increase in an irregular way and we shall take the changes over the second half of the twentieth century to obtain representative average rates of increase in the distances achieved. The results are given in Table 14.2 which lists the average percentage improvement over periods of 10 years.

Event	Average 10 year increase in world record (%)
Hammer	7.8
Discus	5.4
Javelin	8.9
Shot	5.7
Average	**7.0**

We see that the average of the rates of increase is 7% for each 10 years. This is slightly higher than the 6% rate for golf drives over the last 25 years but is somewhat lower than the recent 10% rate. However, it seems quite plausible that the dominant contribution to the improvement in driving performance could be the improved skill and strength of the players.

15 ECONOMICS

The amount of money involved in the professional game is impressive. Many players make millions of dollars a year and the total amount awarded in prizes each year runs into hundreds of millions of dollars. However, this is dwarfed by the amount of money spent by amateurs in payments to golf clubs and the purchase of equipment. No precise statistics are available but we can make a rough estimate of the expenditure on golf worldwide if we take the average spend of, say, 30 million golfers to be 2000 dollars a year, giving a total annual spend of 60 billion dollars. Most of this expenditure goes to golf clubs and we shall start our examination of the economics of golf by looking at the growth in the number of clubs up to the present day figure of more than 30,000.

Golf clubs

In 1744 'several gentlemen of honour' successfully petitioned the City of Edinburgh to provide a
trophy for annual competition on the Links of Leith.
With the trophy provided, the competition went ahead
and in 1764 the Captains of Golf—the previous winners
of the competition—decided 'to admit such Noblemen
and Gentlemen as they approve to be Members of the
Company of Golfers'. With that we have the first golf
club. The Blackheath Golf Club was formed in England
shortly afterwards.

In the United States the South Carolina Golf Club
was formed in 1786. The game then seems to have lost
its appeal until the modern era, American golf restart-
ing when the Oakhurst Golf club was formed in West
Virginia in 1884.

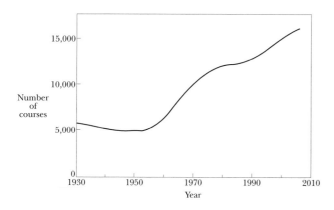

Fig. 15.1. The growth of the number
of golf courses in the United States.

There has since been an almost continual growth in
the number of clubs in the United States, with a substantial expansion in the twentieth century. The available statistics count the number of golf facilities—a

facility sometimes offering more than one course. However, because the term facility is unfamiliar we shall here accept the slight misrepresentation and refer to facilities as golf courses. Figure 15.1 gives a graph of the number of courses from 1931 to the present. Over the last 10 years the United States has, on average, acquired a new golf course every 2 days.

At first sight this rapid growth in the number of golf clubs might be thought to be due to increased affluence, with more people able to afford the game. However, the interpretation is not so straightforward. The population of the United States has grown very rapidly and when we take account of this by calculating the number of golf clubs per capita we find a different story. Figure 15.2 gives a graph of the number of golf courses per 100,000 of the population. We see that the number of courses has just kept up with the growth in the population.

Fig. 15.2. The number of golf courses per 100,000 people in the United States.

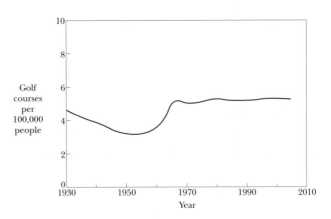

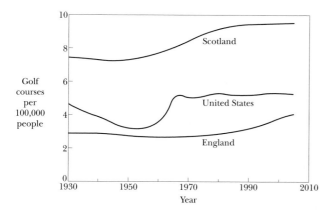

Fig. 15.3. The changes in the number of golf courses per 100,000 people in England, Scotland, and the United States.

Figure 15.3 compares this result for the United States with the experience of both England and Scotland. Over the last 25 years England has seen a growth of one-third in the number of clubs per capita, partly due to the transfer of land usage from farming due to economic factors. Scotland has had a slower growth over the last half century but, having a stronger commitment to the game, the Scots have always had a substantially higher availability of courses.

Golf club economics

An important factor in the economics of a golf club is the proximity of competing clubs. The natural reluctance of golfers to travel unnecessarily long distances means that clubs are competing with all the clubs within a certain range. In England the average density of clubs is very high, about one club per 27 square miles. There is, of course, a considerable variation in the situation of clubs but typically a club will have about 10 competing clubs within a 10-mile radius. It will have an 80% chance that there is another club within 4 miles and an even chance that there are two.

The main income of clubs comes from membership fees and course fees. The relative proportions vary considerably from club to club but it is not unusual for the course fee income to be the larger.

Each golf club has to adjust its fees to ensure that its income exceeds its expenditure. For most clubs this requires careful judgement to ensure that the fees are not so high that they lose custom or too low to provide the required income. Figure 15.4 shows schematically the choice to be made if the income from membership fees is to be maximized, the demand for membership diminishing with increased membership fees. A similar graph will apply for the choice of green fees.

Fig. 15.4. The graph shows how the optimization of income from membership fees depends on the fee.

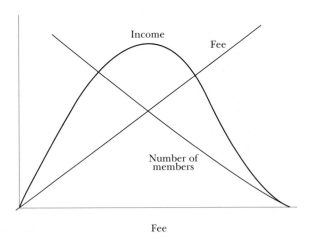

The green fee that is charged depends on the particular circumstances of each club but there is a tendency for clubs with a larger membership to charge higher course fees as shown by Figure 15.5. This is partly because the demand is greater for play at more popular clubs but it also reflects the need to keep the course readily available for the larger number of members.

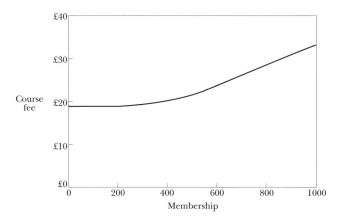

The two basic costs for clubs are administration and course maintenance and usually they are of comparable amounts. The administration costs are similar to those of other organizations—staff, rents, office expenses, depreciation, and so on. The course maintenance involves costs which are specific to golf courses.

There is a requirement to keep the tees, fairways, and greens in a good condition on a daily basis. This calls for the application of a range of fertilizers and weedkillers, and a regular mowing which typically calls for about 1500 hours of mowing each year.

The greens and fairways also need special treatment to provide aeration of the soil. If the ground is allowed to become compacted the roots of the grass find it difficult to penetrate the soil. Aeration is achieved by tining, a process that involves piercing the turf with spikes or removing small plugs of soil—hollow tining. The tining of all the greens on the course will require around a million piercings of the turf. Depending on the location of the course there is a need to drain off excess water in the soil and this is done using longer spikes to provide deep channels that allow the necessary drainage. Insufficient attention to these tasks reduces the quality of the course and leads to a diminished income.

The professional game

It is understandable that professional golfers want to maximize their incomes from the game. This is achieved through the professional associations that represent their interests. The principal vehicle for providing the income is the television coverage of the tournaments organized by the associations. This in turn attracts income from commercial sponsors providing payment of prizes for the tournaments.

Throughout the world there are many professional tours but the professional game is dominated by the PGA Tours of the United States and Europe. The US PGA Tour has a total revenue of over 400 million dollars per annum, the total players' prize money currently amounting to about $260 million. The top players in the Tour each season retain their 'cards', allowing them to play in the Tour again the following season. On the 2006 Tour 125 players retained their cards, all of them winning at least $660,000.

Figure 15.6 gives a graph on a logarithmic scale showing how the total annual prize money on the PGA Tour and the Ladies PGA Tour has grown over the years. A projection of the PGA Tour prize money suggests that it might exceed $1 billion by 2020. The Ladies prize money started at about 10% of the men's and for many years drew closer. In recent years, although the increase in the Ladies prize money has continued, it has not kept pace with the increase in the men's.

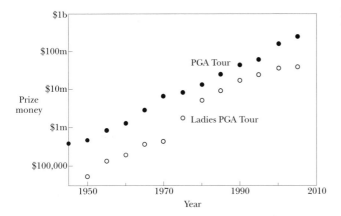

Fig. 15.6. Annual prize money on the US PGA and the Ladies PGA Tours.

Figure 15.6 shows that the present annual prize money on both Tours is more than 500 times larger than that in 1950. However, this exaggerates the increase because there has been a more than 10-fold fall in the real value of the dollar over that period. Figure 15.7 plots a recalculation of the annual prize money on the PGA Tour, giving the value in terms of 2007 dollars. This brings out clearly the dramatic increase in prize money over the last 10 years.

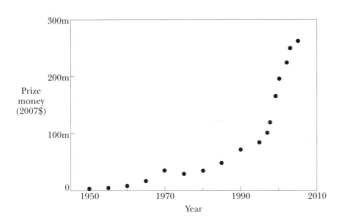

Fig. 15.7. The increase in the annual prize money on the US PGA Tour in real terms.

The European PGA Tour

The European PGA Tour operates in a similar way to the US Tour, but with a total annual prize money of about 120 million euros ($160 million), which is approximately two-thirds of that on the US Tour. In 2006, 118 players kept their cards for the following season and all of them received prize money of more than 200,000 euros ($270,000).

We can obtain an insight into the economics of professional golf by looking at a breakdown of the European Tour's income. Figure 15.8 shows the percentage of the Tour's income from its various sources. It is seen that the largest component is the 54% paid by the promoters to provide the tournament prizes. Television income provides about half as much as the sponsorship. The 9% from event staging consists of contributions such as ticket sales and hospitality.

Fig. 15.8. The composition of the income on the European PGA Tour. *Source*: PGA European Tour.

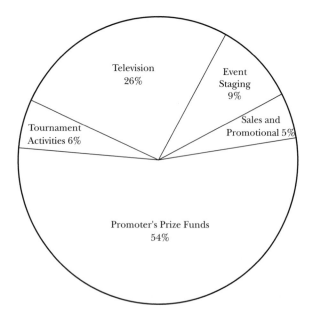

This breakdown illustrates the complex interactions which provide the players with their incomes, the promoters providing prizes in payment for television exposure and the television companies needing entertaining golf from the players to maintain their viewing figures.

BIBLIOGRAPHY

An excellent introductory book is *The Physics of Ball Games* by C.B. Daish (Hodder and Stoughton, 1981). It deals with several ball games but has a strong emphasis on golf.

A. Cochran and I. Stobbs have provided a very good description of the scientific principles involved in the golf swing in *Search for the Perfect Swing* (Triumph Books, 2005).

Dave Pelz has written two outstanding books: *Dave Pelz's Short Game Bible* (Aurum Press, 1999) and *Dave Pelz's Putting Bible* (Aurum Press, 2002). These books combine Pelz's own deep knowledge of the game with an understanding of the scientific basis.

The Physics of Golf (Springer-Verlag, 1993) by Theodore Jorgensen is somewhat more technical than the present book. It covers a range of subjects but concentrates on the swing.

Raymond Penner provides a comprehensive review of the subject in his paper 'The physics of golf', *Reports on Progress in Physics* 66 (2003), p. 131.

In addition, there are the proceedings of the meetings of the World Scientific Congress of Golf entitled *Science and Golf*. These proceedings, I(1990) edited by A.J. Cochran, II(1994), III(1998) edited by A.J. Cochran and M.R. Farrally, and IV(2002) edited by Eric Thain, include a wide range of interesting papers.

Regarding clubs, a book with a lot of technical information is *How golf clubs really work and how to optimize their designs* by Frank Werner and Richard Greig (2000), and an important paper is 'The role of the shaft in the golf swing' by Ronald Milne and John Davis in the *Journal of Biomechanics* 25 (1992), p. 975.

A History of Golf by Robert Browning (A & C Black, 1955) gives a very good insight into the early development of golf.

Newton's laws of motion were proclaimed in his *Mathematical Principles of Natural Philosophy* (London, 1687), usually called *The Principia*. An excellent modern translation and guide with this name is given by I. Bernard Cohen and Anne Whitman (University of California, 1999). Newton's reference to the effect of spin on the flight of a tennis ball was in the *Philosophical Transactions of the Royal Society of London* (1672).

Benjamin Robins's account of the effect of spin on the flight of musket balls is given in his *New Principles of Gunnery* (1742), which was republished in 1972 by The Richmond Publishing Company.

Gustav Magnus reported his research on the force on a rotating cylinder mounted in an airflow in his paper 'On the deviation of projectiles, and on a remarkable phenomenon of rotating bodies' published in the *Memoirs of the Berlin Academy* in 1852 and in an English translation in 1853.

For those with a further interest in fluid dynamics, the classic text on boundary layers is *Boundary Layer Theory* by Hermann Schlichting, first published in German in 1951 and then in English by McGraw-Hill in 1960.

There are many books on general fluid dynamics: a
clear modern text is *Fundamentals of Fluid Mechanics*
by Munson, Young, and Okiishi (Wiley, 1998).

NUMBERS

1 inch	2.54 centimetre		1 centimetre	0.394 inch
1 yard	0.914 metre		1 metre	1.094 yard
1 mile	1.609 kilometre		1 kilometre	0.621 mile
1 mile per hour	0.447 metres per second		1 metre per second	2.24 miles per hour

1 pound	0.454 kilogram			
1 ounce	28.35 gram			
1 pound weight	4.45 newton		1 kilogram	2.20 pound
1 ounce weight	0.278 newton		1 gram	0.0353 ounce
			1 newton	0.225 pound weight
$\pi = 3.14159\ldots$			1 newton	3.60 ounce weight

Golf ball:

Maximum allowed weight	1.620 ounces (45.93 grams)
Minimum allowed diameter	1.68 inches (4.267 centimetres)
Circumference	5.28 inches (13.4 centimetres)
Cross-sectional area	2.21 square inches (14.3 square centimetres)
Volume	2.48 cubic inches (40.7 cubic centimetres)

Diameter of hole	4.25 inches (10.8 centimetres)
Ratio of hole diameter to ball diameter	2.53
Density of air at 20°C (68°F)	1.20 kilograms per cubic metre
	1.20 ounces per cubic foot
Viscosity (kinematic) of air	1.5×10^{-5} m²/s

INDEX